W9-BSM-358

Organize
Your Digital
Life

Also by Aimee Baldridge

The Camera Phone Book

Organize Your Digital Life

How to Store Your Photographs, Music, Videos, & Personal Documents in a Digital World

AIMEE BALDRIDGE

WASHINGTON, D.C.

Table of Contents

Introduction

When it comes to our collections of photos, music, and movies, the conventional wisdom that change is the only constant applies. Sometimes it seems like one minute we were flipping vinyl discs and leafing through leather-bound albums, and the next we were mashing up multimedia files to stream to someone's cell phone. Thanks to all the rapid innovations, we've been given new opportunities to play, share, and publish the media we create and purchase, to transform our older media into digital files, to be more creative with our media than analog formats made possible—and to make a big digital mess. Many of us have seized these opportunities, in the last case by dumping large numbers of mysteriously named files into random places on computers, hard drives, and other tech paraphernalia. And while we accumulate haystacks of new digital media, the records and CDs and photos and films that we once treated so nicely sit moldering in drawers and boxes.

In the midst of all this change, there's one thing that remains the same: messes are easy to make and a pain to clean up. In fact, with technology changing constantly, just keeping up with the way all your gizmos and software work can make getting your digital life in shape seem like an incomprehensible hassle. But it doesn't have to be. What you need to organize your digital life are the same things you need to organize everything else: to know what the organizational tools available are, to be realistic about which ones you will use consistently, and to put them to work. Keeping your digital media organized and safe from disasters isn't about how adept with or interested in technology you are. If you have a computer and digital photos, videos, or music in the first place, you can handle the tools available to manage them.

If the prospect of getting organized isn't enough of a carrot for you, consider the stick: Your hard drive will crash. Discs will become unreadable. It's all just a matter of time. Not only that, but your analog media won't last forever either: Photographic prints and records and tapes and films are all destined to deteriorate eventually, especially if they haven't been preserved in the best conditions. As much as we all love the finger-crossing approach to media preservation, it has proven to be less than effective.

Happily for all of us, digital technology has advanced enough to provide plenty of tools for preserving and organizing all of our personal media, both new and old. And although using them isn't quite as quick and easy as crossing your fingers, many of them are probably a lot easier to use than you might expect. This book is designed to show you what the best tools are and how to use them effectively, so that you can spend your time enjoying and sharing your photos, music, and movies, instead of hunting for needles in your digital haystack.

CHAPTER I

Building a Home for Your Digital Media

Introduction:

Having your photos, videos, and music available in digital form lets you do lots of things that were difficult or impossible to do back in the analog days. But before you can enjoy all those things, a little planning is in order. You need to figure out which of your older media you want to digitize, and where you're going to store everything once you've integrated it into your digital media collection. Where and how you keep your digital media files will affect what you can do with them, how you organize them, and how long they will last.

Choosing the right storage system and buying the equipment necessary to implement your plan is a matter of understanding the costs, capabilities, ease of use, and limitations of each option. The good news is that you don't have to get a computer science degree to make good decisions about your digital media setup.

In this chapter, you'll get a rundown of all your storage options, as well as tools to help you assess your entire media collection and make a plan to bring it together in digital form. You'll also get some tips on how to protect your digital collection from disasters and recover it if the worst happens.

Assessing Your Media Collection

The Goal:

Take an inventory of the physical media items that you want in your media collection.

What You Need:

Copies of the **Media Collection Inventory List** to fill out on paper or a laptop.

Time Required:

About ten minutes per room.

Before creating a system for organizing your media collection, take an inventory of all of the physical media items you would like to include in it. The goal is to get an overview that will help you prioritize tasks and to determine the kinds of archiving tools and amount of storage space you will need. The **Media Collection Inventory** ▸ can help. Make one copy of it for every room in your home, or re-create it in word-processing or spreadsheet software if you're comfortable with that. Then take it on paper or on a laptop to each room in your home and fill it out, using a separate list for each room. When you're done, compile the results into a single master list.

For this step, leave the Priority Score and Space Estimate columns blank. You'll fill them in later.

Tips for Efficient Tallying

✔ **Count, don't browse.** If you start looking over things, sizing up your media collection will take all day. To prevent yourself from browsing, set a timer or an alarm and allow ten minutes per room. If time runs out before you're done, reset the clock and finish the room you're in, but try to beat the clock each time you set it. If you catch yourself browsing, make a little star next to the item on your list to indicate it's something you want to come back to after you're done with your tally.

✔ **Multiply whenever possible.** Estimate the number of photographic prints in a box or records on a shelf by counting how many there are in an inch-high stack, then multiplying that number by the total height of the whole stack of prints or records. For photo albums that generally have the same number of prints per page, multiply that number by the number of pages.

✔ **Estimate lengths.** Don't spend time trying to ascertain the exact duration of video and audio recordings on tape at this point. Just note the full recordable length of the cassettes.

▸▸ Media Collection Inventory

Photographs

CATEGORY	#	FORMAT	SPACE ESTIMATE	PRIORITY SCORE	DESCRIPTION
Small prints	24	Wallet-size			School portraits
4x6 prints	200				Holiday and vacation photos
5x7 prints	16				School portraits
Large prints (8x10+)	20	15 8x10s, 5 11x14s			Framed family and vacation photos
35mm slides	120				Vacation photos
35mm negatives	240				Holiday and vacation photos
Other	48	110 format negatives			Parties and home snapshots

Home Movies

CATEGORY	#	FORMAT	SPACE ESTIMATE	PRIORITY SCORE	DESCRIPTION
Films	2	50″, Super8			Creative film projects
VHS & Betamax	12	6 hours, VHS			School plays
8mm & Hi8	6	120 min., Hi8			Vacation videos
MiniDV	14	60 min.			Holidays and school events
DVDs & Video CDs	5	30 min., mini DVD-Rs			Sports events
Other	2	60 min., MicroMVs			Vacation videos
Other	48	110 format negatives			Parties and home snapshots

Music

CATEGORY	#	FORMAT	SPACE ESTIMATE	PRIORITY SCORE	DESCRIPTION
Vinyl records	70	60 LPs, 10 singles			Jazz and folk
Cassette tapes	26	All 60 min.			80s pop
8-track tapes	9				70s disco
CDs	225				Various genres
Other					

Media Collection Inventory

For you to fill out yourself.

Photographs

CATEGORY	NUMBER	FORMAT
Small prints		
4x6 prints		
5x7 prints		
Large prints (8x10+)		
35mm slides		
35mm negatives		
Other		

Home Movies

CATEGORY	NUMBER	FORMAT
Films		
VHS & Betamax		
8mm & Hi8		
MiniDV		
DVDs & Video CDs		
Other		

Music

CATEGORY	NUMBER	FORMAT
Vinyl Records		
Cassette Tapes		
8-Track tapes		
CDs		
Other		

SPACE ESTIMATE	PRIORITY SCORE	DESCRIPTION

SPACE ESTIMATE	PRIORITY SCORE	DESCRIPTION

SPACE ESTIMATE	PRIORITY SCORE	DESCRIPTION

Deciding What to Digitize

There are many advantages to bringing your analog media into the digital realm. But the transfer does require time and money to accomplish. Once you've taken a tally of your analog collection, make some decisions about what you want to digitize and what just isn't worth the trouble. While you're at it, consider whether the items that aren't worth the trouble are worth keeping at all. You may find that you can clear out some shelf space by giving away or selling the items that don't make the cut.

Fifteen Questions to Help You Make Up Your Mind

Consider the items you listed in your Media Collection Inventory. Ask yourself the following questions about each item to decide whether it's worth digitizing. If you answer Yes to any of the questions below, then keep the item on the list. If not, cross it off.

✔ **1. Is your personal collection the only place** where the media can be found? If your media are available from Web sites or services, it may not be worth the time and money required to create and store digital copies of them. That is often the case with music and movies, whereas photo collections are usually unique and irreplaceable.

✔ **2. Do you want to share your media with others?** In digital form, the content can be shared with multiple parties, or even the public, over long distances.

✔ **3. Do you want to carry this content on a portable device?** If your photos, music, or even videos are in digital form, you can carry hundreds or thousands of them with you, or store them on a Web site that is accessible via a mobile device.

✔ **4. Do you want to restore the content** to its original quality? Software can work wonders with faded and damaged photographs and scratchy records. Once your content is in digital form, you can make new prints or recordings that will outshine the aging originals.

✔ **5. Do you want to use the media in creative projects?** Once you have digital copies of your media, the possibility for doing creative things with the content is seemingly endless. You can create multimedia shows; apply different looks and artistic effects to photos; add soundtracks to home movies; make posters, calendars, or gift items out of your pictures; or just make new prints for a scrapbook or framing.

✔ **6. Is it difficult to find the things you want to play or display** in your analog collections? Sometimes it's worthwhile to convert your media to digital form just because the contents are easier to locate. If you never listen to old records or look at old photos because you don't want to spend half an hour hunting through your collections, you might be able to enjoy them again by making them instantly accessible through a digital player or software.

✔ **7. Do you want to add descriptive information?** When your content is digitized, it's easy to attach captions, historical notes, and other descriptive information to it. Read more about adding information to photos on page 84. When you create digital copies of your music and movies, you may also be able to attach publicly available information to them about the artist, album, and other elements.

✔ **8. Do you want to pass your media down to future generations?** Converting your media to digital form can help you preserve the contents and ensure that they will still be playable when older devices such as turntables and tape players are no longer widely available.

✔ **9. Do you want to use the content of your analog media on a Web site** or online social network? If so, this decision is a no-brainer—you have to digitize.

✔ **10. Do you want to be able to play or display your content at special events?** Using digital players at a wedding, reunion, or other event can be easier and more convenient than employing older technology such as record players, slide projectors, and video cassette players. When your media are digital, you can create a playlist or slideshow that lasts for hours and doesn't require attention after you hit Play.

✔ **11. Are the original materials in poor condition** or likely to deteriorate in the near future? Creating digital copies may allow you to preserve and restore content that will be lost when the paper, vinyl, or tape it's stored on deteriorates beyond repair.

✔ **12. Are you planning to donate the originals** to an institution or give them to another person? Creating digital versions grants you the option of giving your media away while retaining copies for yourself.

✔ **13. Have you lost the devices needed to play or view your media?** If you don't have the record or cassette players you need to play your tapes and records, or if the devices are broken, you can have digital copies made by a service or buy a digitizing device to convert them yourself.

✔ **14. Has the company that made the devices** needed to play your media stopped selling or supporting them? This question is especially relevant to audio and video cassettes. As companies stop making players for the formats you own, it becomes more difficult and expensive to have your existing equipment repaired or replaced.

✔ **15. Are you willing to spend a little money** to create digital copies of your media? It's likely that you will have to pay something to digitize and store your media. Depending on the resources you already have, the nature of your collection, and the digital storage options you choose, your expenses could fall below $100 or run closer to $1000. The next section helps you to estimate the amount of storage space you'll need so that you can get an idea of what it will cost to create and maintain your digital media. If it's clear to you off the bat that you're not willing or able to spend any money at all on digitizing a particular item, then it's probably a good candidate to cross off your list.

Assessing Your Media Organization Tools

In each of the chapters in this book, you'll find information about tools you can use to organize and manage your digital media collection. Before you buy new software, make sure to take a careful look at the programs you already have to see if they offer the tools and features you need. Here are a few places to check:

Your operating system. Throughout this book, there is information about where to find the tools that are built into the Windows or Mac operating system of your computer.

On discs included with hardware. Many computer-related devices, such as hard drives and scanners, come with software packages that do more than just make the hardware work. Consumer electronics such as digital cameras and MP3 players almost always come with software for managing and editing media. Check the boxes you might have left discs in and the drawers where you might have tossed them without looking at their contents to see what the software programs that you might have overlooked can do.

On Web sites where you have an account. Web sites related to digital media, such as online image galleries, sometimes provide software that you can download or offer online features that are useful for managing digital media. See what the latest tools available from sites where you have a subscription or an account are.

If you want to expand your search for software tools beyond the resources described in this book, go to the mother lode of downloadable software, Download.com. The site offers a vast collection of software download links, with brief descriptions, user ratings, and, in some cases, editorial reviews of the programs. If you're on a budget and are looking for free options, this is a good place to hunt for them.

You should also make sure you're getting the best performance out of what you already own. Both software and hardware makers frequently make improvements to their products that are available for free or at a low cost to current owners. Here are some ways to take advantage of this, and to make sure you have adequate basic connectivity to external devices and Internet services:

Upgrade your operating system. The latest versions of both Windows and Mac operating systems provide the best tools for organizing digital media, backing up files, and integrating desktop features with online services. Make sure your computer meets the system requirements of the new operating system before you upgrade.

Upgrade your software. If the programs you use to manage your digital media don't have a "check for updates" option in their help menu, go to the software makers' Web sites to see if there are upgrades available.

Upgrade your firmware and drivers. Components of your computer such as optical drives rely on software called firmware and drivers to run. The latest versions are likely to improve the performance of your hardware, and possibly the quality of the media you create with it. Check the support sections of the manufacturers' Web sites for downloadable updates. You can view the current firmware and driver versions of your components in Windows by opening the Control Panel and going to System/Hardware/Device Manager. Right click on a component and select Properties/Driver. On a Mac, click the Apple logo in the upper left corner of your screen and select About This Mac to see information about your computer's components.

Note about Windows Explorer ▸▸
In many places, this book refers to the Windows Explorer file browser. It's easy to confuse this with Microsoft's Internet Explorer Web browser because the names are so similar, but they are two different programs. Windows Explorer is the program that lets you browse all of your files and folders. You can open it by clicking on Computer or My Computer.

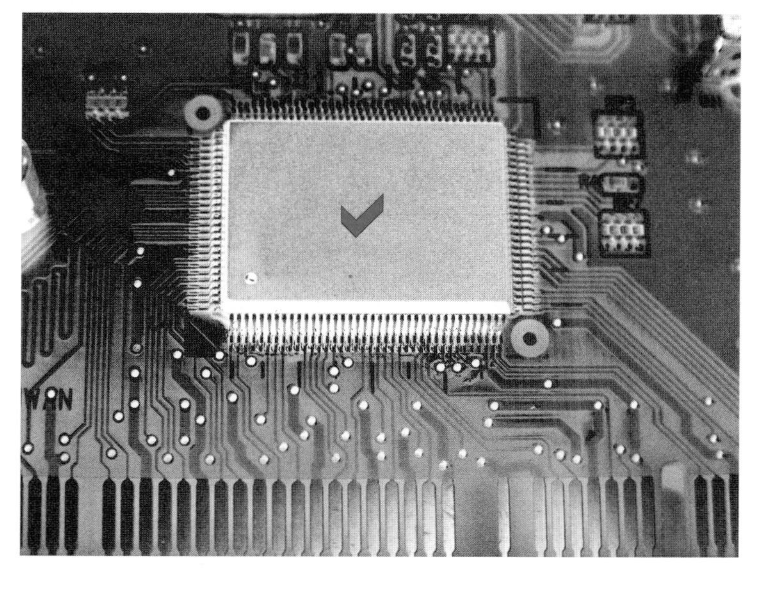

Check your USB ports. If you have an older computer, make sure the ports support USB 2.0, and not the outdated USB 1.1 standard. Look at your computer's documentation to check this. USB 2.0 gives you faster performance from attached devices, which can be important if you're doing something like archiving your media on an external drive. If your computer only has USB 1.1, you may be able to upgrade it by adding an inexpensive **PCI or PCI Express USB 2.0 card.** ▸ To do so, you have to open the computer case. Make sure you turn off the power supply on the back of your computer and touch something metal to discharge any static electricity you may be harboring before you reach into your machine and start tinkering. Then install the new card in an empty PCI or PCI Express slot on the motherboard so that the card's USB ports are available on the back of the computer.

Get a faster Internet connection. If you're still using dial-up or have a DSL connection that just isn't very speedy, take a look at what your Internet service provider is offering in the way of faster DSL or a cable connection. Having an adequate connection is necessary to use many online tools and services.

And don't forget to use antivirus software to protect your system—and the media collection stored on it—from attacks.

Estimating Your Storage Needs

There are numerous factors that contribute to the size of a digital media file, including file formats, bit rates, resolution, and even how detailed the content is. You can use the information provided here to estimate how much storage space you'll need to store the digital files you create when you rip discs and digitize your analog photos, videos, and music. Multiply the approximate size information by the number of items you've entered on your Media Collection Inventory. Divide the total number of megabytes by 1000 to see how many gigabytes of storage you'll need and enter that number in the Space Estimate column.

Making an estimate with the information provided here won't tell you exactly how much storage space you'll need for your files once you've digitized your analog media. However, you can use it to get a general idea of the amount of space your files will require. Doing this prevents you from seriously underestimating your storage requirements, and gives you an idea of how much you'll need to spend on storage devices. To make more accurate estimates, do test scans or recordings of your analog media with the settings you plan to use.

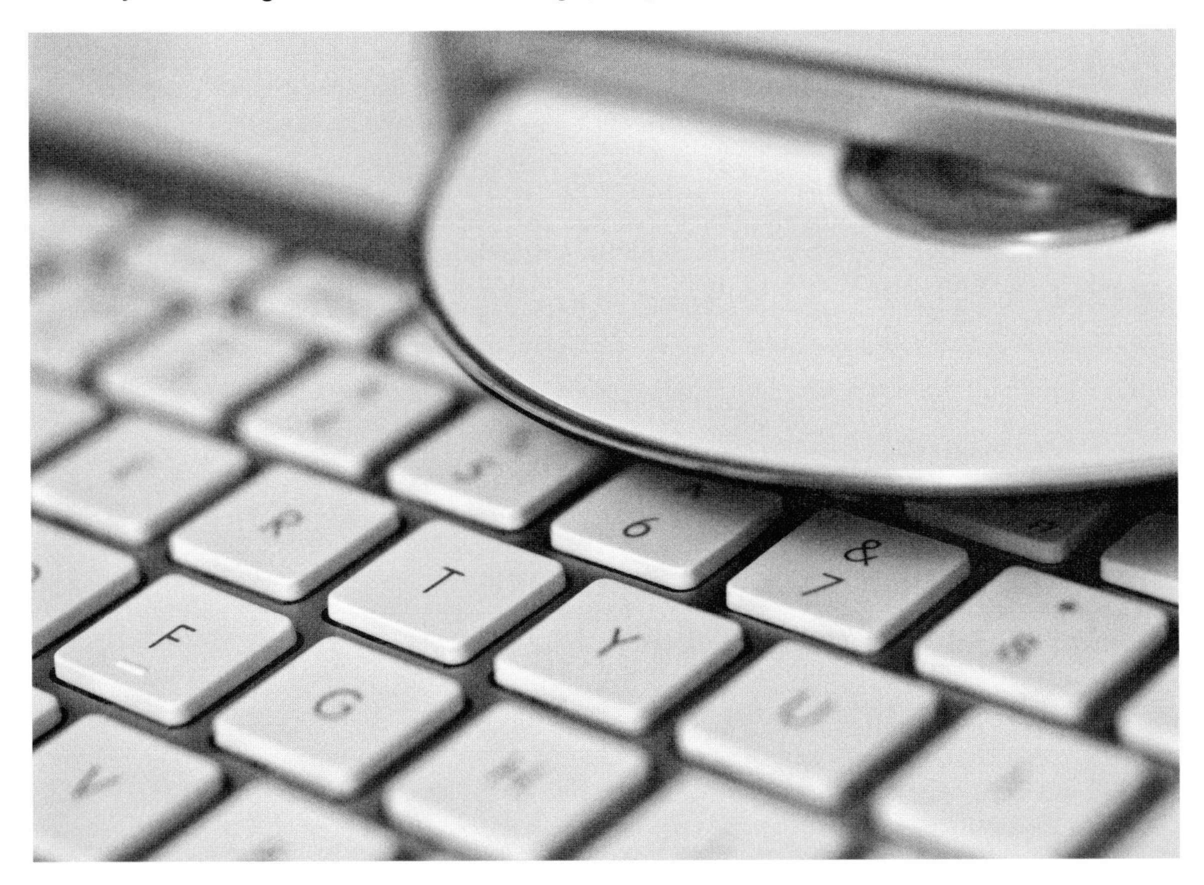

	ORIGINAL MEDIA	SIZE/LENGTH	DIGITAL FILE DETAILS	APPROXIMATE SIZE
PHOTOGRAPHY	4x6 prints	1	300 dpi scanning resolution, 8-bit JPEG	1MB
		1	300 dpi scanning resolution, 8-bit TIFF	6MB
		1	300 dpi scanning resolution, 16-bit TIFF	12MB
	5x7 prints	1	300 dpi scanning resolution, 8-bit JPEG	1.5MB
		1	300 dpi scanning resolution, 8-bit TIFF	9MB
		1	300 dpi scanning resolution, 16-bit TIFF	17MB
	8x10 prints	1	300 dpi scanning resolution, 8-bit JPEG	2MB
		1	300 dpi scanning resolution, 8-bit TIFF	22MB
		1	300 dpi scanning resolution, 16-bit TIFF	40MB
	35mm slides or negatives	1	2000 dpi scanning resolution, grayscale scan, 8-bit JPEG	1.5MB
		1	2000 dpi scanning resolution, grayscale scan, 8-bit TIFF	5MB
		1	2000 dpi scanning resolution, grayscale scan, 16-bit TIFF	10MB
		1	2000 dpi scanning resolution, 24-bit color scan, 8-bit JPEG	2MB
		1	2000 dpi scanning resolution, 24-bit color scan, 8-bit TIFF	15MB
		1	2000 dpi scanning resolution, 24-bit color scan, 16-bit TIFF	30MB
FILM, VIDEO	Films, video cassettes, DVDs	1 hour	Uncompressed	95GB
		1 hour	DV-AVI on MiniDV	12GB
		1 hour	MPEG-2 on DVD	4.7GB
MUSIC	Vinyl records (LPs)	1	WAV	450MB
		1	MP3 (128kbps)	45MB
	Cassette tapes	1	WAV	450MB
		1	MP3 (128kbps)	45MB
	CDs	1	WAV	600MB
		1	MP3 (128kbps)	60MB

Choosing Hard Drives and Storage Devices

Once you've figured out roughly how much space you'll need to store and back up your media and other files, you can select a storage device or devices. Take these factors into consideration:

✔ **1. Capacity limit.** Again, how much storage do you need, now and in the near future? Find out how to make an estimate on page 18. Some storage device types can be purchased in or expanded to capacities of several terabytes, while others have much lower limits.

✔ **2. Redundancy.** A storage device that offers redundancy keeps more than one copy of each of your files on separate hard drives. If one drive fails, your files won't be lost. This is especially important if you're using your storage device as a media archive and don't have duplicate files stored elsewhere. Multidrive systems that offer redundancy include RAID 1, RAID 5, a Windows Home Server, or a Drobo system.

✔ **3. Connectivity.** Do you want your storage device to be available for just one computer or for multiple computers and other devices? If you have a home network, you can use network-attached storage or a home server that all of your computers can back up to and share media on. If you will be connecting your storage device directly to one or more computers, make sure they have the necessary port type for the device you buy. External hard drives and enclosures with more than one connection type are available.

✔ **4. Portability.** Do you want to be able to carry your storage device between home and office, or on trips, or to leave it in a safe deposit box? Do you want to back up two computers in different locations with a single device? Having a small, lightweight device that does not require any installation process when connected to a new computer (in other words, one that is "plug-and-play") may work best for you.

✔ **5. Backup frequency.** How often do you want to back up? The safest and easiest way to back up is to set up a continuous automatic backup. For that to work, you'll need a storage device that is always attached to your computer or network.

✔ **6. Tech savvy required.** How comfortable are you with technology? Most of the options described here are extremely easy to set up, but some require a little bit of tech savvy and the willingness to go through a more detailed setup process or even loosen a few screws. Stick with something that you feel confident about setting up and using, so that you don't put off doing it.

✔ **7. Durability.** If you need to carry your backup or archive device with you, look for one that is built to resist damage. If you live or travel in areas prone to extreme temperatures and humidity, you should avoid storing your files on media that are susceptible to damage from those conditions.

✔ **8. Cost.** Cost will depend on the capacity you need, but some storage device types have higher starting prices and higher costs per gigabyte (GB).

✔ **9. Media sharing capabilities.** If you want to play your media collection on your TV or stereo, look for devices that will make your files available either through a direct connection or over a network. Some external hard drives can connect directly to a TV or stereo to play music, videos, and photos. Network-attached storage devices that support the UPnP AV standard can stream media to a network media player attached to a TV or to a network-attached stereo. If you'll be transferring or streaming video files, look for a fast gigabit Ethernet connection in a networked storage device.

✔ **10. Included software.** If you don't already have backup, sync, or drive imaging software, buying a storage device that comes with it can be a good deal. External hard drives and enclosures and network-attached storage usually come with a bundle of software. Flash drives and cards sometimes come with file-recovery software.

Storage Device Types

Internal hard drive. An internal hard drive is an additional hard drive installed in your desktop computer. Current computers use Serial Advanced Technology Attachment (SATA) hard drives. Older machines use ATA drives, sometimes called IDE, PATA, or EIDE. Read about how to install an internal drive on page 22.

External hard drive. An external hard drive attaches to your computer via a USB 2.0, FireWire 400 or 800, or eSATA connection. External drives range from highly portable palm-size models to larger, heavier models. Look for one-touch backup, which lets you launch backup software with the touch of a button.

External hard drive enclosure. External hard drive enclosures ▶ can house two or more hard drives, and connect to your computer the same way as a single external hard drive. They usually offer some type of redundancy, such as RAID or a proprietary system. (Data Robotics's Drobo is an example of the latter.) Some are also compatible with accessories that convert them into network-attached storage devices. Most come with backup software. Some enclosures can be purchased with hard drives installed; others are purchased without drives, and you buy internal SATA hard drives separately. Look for hot-swappable drives, which can be easily removed and replaced.

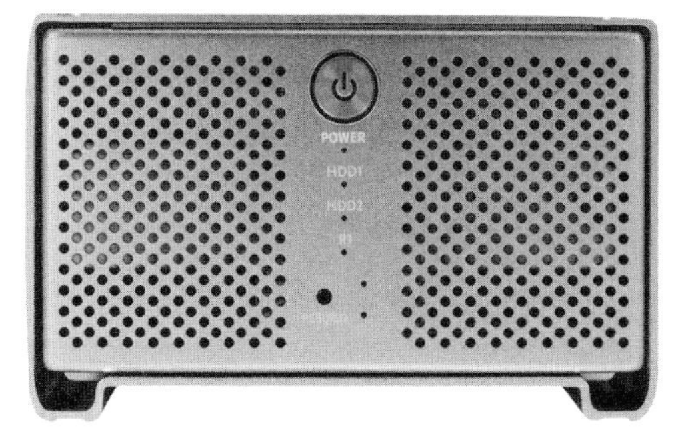

Network-attached storage (NAS). NAS devices are like little computers that include a controller and one or more hard drives. When you attach a NAS device to your wired or wireless home network, it becomes accessible to all of your computers and other networked devices. And as long as you leave it on, it can also make all of the files stored on it available from any Internet-connected computer, since it functions as a server. You can buy a NAS enclosure with drives preinstalled or install your own. The installation process can be as simple as sliding a drive into a slot. Look for features such as UPnP AV support for streaming media to networked media players and stereos, a USB port for connecting a printer or other device to make it accessible to your whole home network, and included software tools for hosting a Web site and accessing files over the Internet. Mac users can benefit from the software integration that Apple's Time Capsule NAS device offers.

Windows Home Server. As the name suggests, a Windows Home Server is a computer whose purpose is to store all of your digital files and make them available to other computers and devices on a home network. It automatically backs up all connected computers and makes their contents accessible online too.

USB flash drive, memory card, or portable hard drive device. These little devices are the most portable. They're small enough to carry on a keychain or even in a wallet. Many people use them simply to transfer files easily between computers and other devices, but they can also be handy for backing up or archiving a selection of important files. Their portability makes them easy to store in a safe deposit box or a small home safe. In addition to USB drives and memory cards that use flash memory to store files, portable hard-drive-based devices such as iPods and other media players can be used to store files.

Optical storage (discs). Optical discs include CDs, DVDs, and Blu-ray discs. Read more about using them to store your media on page 32.

Installing an Internal Hard Drive

If you're comfortable tinkering with your PC a little, you may be able to install an internal drive in your system yourself. Here's what you need to do:

✔ **1. Make sure that you are not carrying a static charge** when you touch anything inside your computer. Touch something else that's metal before you start, to discharge any static electricity. Don't work near anything that might give you a charge, such as carpeting under your fuzzy slippers. If your environment is static-prone, you can buy an antistatic wristband from an electronics store and wear it while you work.

✔ **2. Turn your computer off,** turn the power switch on the back of the machine off, and unplug it.

✔ **3. Remove the side panel** of your computer so that you have access to the inside. You will probably have to unscrew pegs that are holding it in place on the back of the computer case.

✔ **4. Look in the front area of the computer,** and locate the currently installed hard drives. Unscrew and pull out the hard drive enclosure if necessary.

✔ **5. Put your new hard drive** in an available space in the enclosure, and secure it with the screws that came with it, in the same manner as the other drives.

✔ **6. Connect one end of the data cable** that came with your drive to the drive, and the other end to the appropriate connector on the motherboard—the main circuit board that all of your computer's components are attached to. If you're not sure where to connect it, get a flashlight and look at the labels next to the existing connectors for a label reading "SATA." You can also probably find a guide to the motherboard on the manufacturer's Web site. The manufacturer's name and a model name will appear on the board.

✔ **7. Look for the cable extending** from the computer's power supply. This is the big box with the switch on the outside that you turned off before you started. Find an unused SATA power connector along the cable, and connect it to your drive.

✔ **8. Put the side panel back on,** plug the computer back in, switch the power on, and turn your computer on. The drive should appear when you look in Windows Explorer.

The process is the same for older computers that use IDE drives, but the connectors look different. However, it's more common to encounter problems installing hardware in older machines; if you don't feel confident about your ability to troubleshoot if something goes wrong, don't do it.

Setting Up Network-Attached Storage

The nerdy-sounding name of this type of storage might lead you to believe that it's complicated to use. But if you already have a home network and buy a NAS device that's designed for consumers and not system administrators, you should be able to have it up and running in about fifteen minutes. You'll need to follow the specific instructions that come with your device, but here is an overview of how the process works in general:

✔ **1. Plug in the NAS** device and turn it on.

✔ **2. Connect it to your home network's** router with an RJ-45 LAN (Ethernet) cable.

✔ **3. Install the included software** on the computers connected to your home network in order to make the NAS device available to them.

✔ **4. In Windows, map the NAS device** as a network drive so that it appears in the Explorer file browser along with your local drives. You can do this easily by going into Explorer and selecting Map Network Drive from the Tools menu.

✔ **5. Set up your backup software** (most NAS devices come with backup software) to back up your networked computers to the NAS device.

Note that some NAS devices have a wireless router built in. If you use this type, you will plug your Internet connection cable directly into the NAS instead of using a separate router.

How Your Storage Options Compare

Use this to figure out which device type or combination of devices best meets your needs.

	CAPACITY LIMIT	REDUNDANCY	CONNECTIVITY	PORTABILITY
INTERNAL HARD DRIVE	High if one or more drive bays are open in your computer	Available in some computers with RAID 1, usually an expensive option (or requires high level of tech savvy to set up yourself)	Open internal drive bay required; SATA or IDE connection	None
EXTERNAL HARD DRIVE	High	Available in more expensive models with multiple drives in one enclosure	USB 2.0, FireWire 400 or 800, or eSATA	Depends on size; plug and play
EXTERNAL HARD DRIVE ENCLOSURE	Very high	Available	USB 2.0, FireWire 400 or 800, or eSATA	None
NETWORK-ATTACHED STORAGE	Very high	Available	Ethernet; connects to home network router	None
FLASH DRIVE, MEMORY CARD, OR PORTABLE HARD DRIVE DEVICE	Small or Moderate	No	USB 2.0 or built-in card reader	Very portable
WINDOWS HOME SERVER	Very High	Available	Ethernet; connects to home network router	None
OPTICAL STORAGE (DISCS)	Small or moderate per disc	No	Insert in optical disc drive of computer or other device	Portable

COST	BACKUPS	TECH SAVVY	DURABILITY	MEDIA SHARING	
Starts under $100; good value per GB	Allows continuous backup	Moderate savvy needed to install drive	Low susceptibility to environmental damage	Depends on other computer hardware and software	**INTERNAL HARD DRIVE**
Starts under $100; good value per GB	Allows continuous backup when left plugged into computer	Very low for single drives	Drives with ruggedized exteriors available, as well as fireproof and waterproof drives	Depends on other computer hardware and software	**EXTERNAL HARD DRIVE**
Starts under $200; good value per GB	Allows continuous backup	Low; enclosures with multiple drives and RAID options may require a little setup	Low susceptibility to environmental damage	Depends on other computer hardware and software	**EXTERNAL HARD DRIVE ENCLOSURE**
Starts under $200; good value per GB	Allows continuous backup	Low for basic setup; moderate for some options	Low susceptibility to environmental damage	Available	**NETWORK-ATTACHED STORAGE**
Starts under $10; high cost per GB	Backs up only when attached	Very low	Ruggedized and waterproof models available	Depends on computer or device it's plugged into	**FLASH DRIVE, MEMORY CARD, OR PORTABLE HARD DRIVE DEVICE**
Starts over $400; good value per GB	Allows continuous backup	Low	Low susceptibility to environmental damage	Available	**WINDOWS HOME SERVER**
Starts under $1 per disc; good value per GB in CDs and DVDs, higher cost in Blu-ray	Manual backup	Low	Susceptible to environmental damage if not stored properly	Depends on computer or device it's inserted in	**OPTICAL STORAGE (DISCS)**

Attention Videophiles!

If you have very large files, such as video files or files generated by some backup software, don't buy a hard drive or NAS device that can only use the FAT32 file system. FAT32 does not support files larger than 4GB. If your drive comes formatted in FAT32, reformat it to NTFS or Mac OS Extended. You can use the NAS software to do this, or, with an individual hard drive, you can do it via your file browser. In Windows Explorer, right click on the drive letter, select Format, then NTFS. In Mac Finder, open Applications/Utilities and click on Disk Utility. Select Erase, then choose Mac OS Extended from the drop-down menu.

Using a Storage Device with RAID

A storage device or computer that includes more than one hard drive set up to function as a Redundant Array of Inexpensive Drives (RAID) can be a good place to store files that you want to preserve, because it offers redundancy. However, for this to be true, you must select the right type of RAID. If you are using a device that has two hard drives, you should select RAID 1, which turns the second drive into a clone of the first, storing and updating identical copies of your files on both drives. If one drive fails, you will be able to retrieve all of your files from the other drive. Of course, RAID 1 cuts the storage capacity of your device as a whole in half. For example, if you purchase a 500GB NAS device with RAID 1, it will provide you with 250GB of storage, since everything on it is being stored twice.

In the case of a RAID 1 drive failure, you should replace the failed drive right away. This is because identical drives are generally used in RAID devices. They are likely to have been manufactured in the same batch, and therefore likely to have similar weaknesses and life spans. So, if one drive goes, that's an indication that its twin may be on its last legs. You should also restore redundancy to the system as soon as possible to ensure that your files are preserved. You will need to replace the failed drive with one that has identical capacity and other specifications. The necessity of the drives in a RAID system

Hard Drive, Enclosure, and NAS Makers ▸▸
Apple www.apple.com
Beyond Micro www.beyondmicro.com
Buffalo Technology www.buffalotech.com
Data Robotics www.drobo.com
D-Link www.dlink.com
Fantom Drives www.fantomdrives.com
Iomega www.iomega.com
Kanguru Solutions www.kanguru.com
LaCie www.lacie.com
Netgear www.netgear.com
Synology www.synology.com
Western Digital www.westerndigital.com

Selected Flash Memory Card and USB Drive Makers ▸▸
ATP www.atpinc.com
Kingston www.kingston.com
Lexar www.lexar.com
PNY www.pny.com
SanDisk www.sandisk.com
SimpleTech www.simpletech.com

being identical can make RAID storage problematic as a long-term archiving tool: If you can't purchase an identical hard drive a few years down the line, then the entire system will have to be replaced instead of just one drive.

If you need a high-capacity storage device and purchase one that has more than two drives, you may have the option of selecting RAID 5. This type of RAID also provides redundancy, but it does so by distributing duplicates of your data over multiple drives. If one drive in a RAID 5 system fails, you will, in principle, be able to recover your files from the remaining drives. I don't recommend using one if you don't have a fairly high level of knowledge about computer technology, however. The process of restoring files can get complicated with RAID5.

Other types of RAID that are generally available to consumers are not suitable for archiving or backup purposes. RAID 0 is a common option in consumer RAID devices, but it does not offer redundancy. Its purpose is to distribute data over more than one hard drive in such a way that system performance is improved. It's useful to people who are doing advanced work with very large files that can slow a computer down, notably video files. However, it has no use as an archiving or backup tool.

RAID is often viewed as a complicated system that only advanced computer users should attempt to use. However, there are many consumer storage devices that come with preconfigured RAID and require no setup other than checking the device's RAID management software to make sure that the right type of RAID is selected. The management software should also make restoring a RAID 1 system straightforward.

Zero Minute Backups

If just seeing the words "hard drive" and "backup software" makes your eyes glaze over, a Clickfree drive may be thing for you. These little hard drives made by Storage Appliance Corporation (www.goclickfree.com) have built-in software that completely automates the process of backing up a Windows computer. All you have to do is plug the drive into a USB port on your system. The Clickfree software starts automatically and backs up all of the media files, documents, and e-mail on your system. The drive is even powered via USB, so you don't have to plug it into a power outlet. You can back up multiple computers on a single Clickfree drive, as long as the amount of data doesn't exceed its capacity. The drive's software also provides full-fledged backup options in a simple interface, in case you'd like to customize your backups by selecting particular folders, file types, or whole drives to save. The drives are small and light enough to carry around or store in a safe deposit box, too.

Creating a Media Storage System

Your storage system should accomplish some basic goals. You should be able to recover your media in the event of a device failure, disaster, or theft. Your system should make media easily accessible for playing and displaying, sharing online and on multiple devices, and editing or including in projects. It should also make it possible to restore the media you use frequently without a lengthy or expensive process. To meet these criteria, you need to have your photographs, videos, and music, stored on or available from three types of sources: accessible home storage; a home backup and archive; and a remote backup and archive. Here is an example of how to set up a comprehensive media storage system.

- **Store all of your media on a high-capacity,** hard-drive-based device with redundancy, such as a computer with RAID, a NAS (network-attached storage) or other redundant hard drive array, a Windows Home Server, or a computer set up to automatically back up to an external hard drive. If you have more than one computer and a home network, a server or NAS makes more sense. Backing up to an external hard drive does the trick with a single computer for a lower cost.
- **Archive your photos** on optical discs periodically, once a month or more or less often, depending on how much you shoot. Store the discs at a remote location.
- **Archive your videos** on digital video cassettes, discs, or a portable hard drive, and store the archived videos at a remote location.
- **Archive your music** on discs or a portable hard drive, and store the discs or drive at a remote location.
- **Back up or sync your media** and important documents to online storage and galleries.
- **Back up your important documents** on a flash memory card or USB drive and store it in a remote location such as a safe deposit box.

The redundant hard-drive storage accomplishes two things: First, it makes your media easily accessible. Second, it allows you to restore your frequently used media quickly in the event of a device failure, simply by replacing a hard drive. The archived media stored in remote locations protects it against loss in the event of a disaster or theft, and the online storage not only provides additional insurance against loss, but also gives you remote access to your media.

If money is no object and your media collection is a manageable size, setting up a system like the one described above will cover all your bases. However, most people have to balance current costs against future risks. In other words, we can't all afford the multi-drive NAS that would be required to store our whole media collection

Multimedia Organizers ▸▸
There are many software options listed in this book for managing specific types of media files, including photos, music, and videos. Often, programs that specialize in a particular media category provide a more powerful set of tools for handling it. However, some people prefer a more centralized approach to media management. These software organizers are designed to manage all three categories of media with one unified program.

ArcSoft MediaImpression
www.arcsoft.com
J. River Media Center
www.jrmediacenter.com
Microsoft Expression Media
www.microsoft.com/expression

on a redundant system at home, or the amount of online storage space it would take to back up everything online. As a result, we have to distinguish between the things that are crucial to devote storage space to and the things that we can take some risks with.

The main issue here is whether there is any way to replace the content *at all* when the device it is stored on can no longer be used. Consider storage failure inevitable. Hard drives crash, optical discs deteriorate, and flash memory develops glitches. It's all just a matter of time. If you want to preserve your irreplaceable content, you must back it up, so that when one copy goes, you can create another duplicate on the latest and greatest storage device. Personal photos, home videos, and home-recorded music should be at the top of your backup list. To help you figure out what to devote your resources to and what to take some risks with, here are some items that you can consider excluding from your backup or archiving system:

- **Music ripped from CDs or recorded from vinyl records.** Treat the original CDs and records as backup copies and store them at a remote location. If the music on your hard drive or optical disc is lost, you'll have to transfer them from CD or vinyl all over again, but the music won't be lost. (Treating audio cassette tapes as a backup is a bad idea, though, since they are likely to deteriorate more quickly.)

- **All commercially recorded music.** It will still be available from commercial sources if you lose your copies, although you will have to pay for it again.

- **All commercially recorded videos.** They will still be available from commercial sources if you lose your copies, although you will have to pay for them again, too.

- **Home videos transferred to a computer** or optical discs from digital video cassettes. Treat the original cassettes as backup copies and store them at a remote location. If the videos on your hard drive or optical discs are lost, you'll have to transfer them from cassette all over again, but the footage won't be lost. Stored properly, the cassettes can be expected to last more than ten years, and possibly several decades.

Prioritizing Your Media Organizing Tasks

The first thing to do, before you get started on organizing your media, is to back up the contents of your computer. Don't put it off! Computer hard drives can crash unexpectedly. If you have an available external hard drive or some other type of storage option, just use the built-in options that your Windows or Mac operating system provides to back up your system to the drive. External hard drives are relatively inexpensive these days, so consider picking one up to back up the media and documents that you currently have on your computer—even before you estimate the additional storage space you'll need to house your whole media collection after you digitize your analog media.

Once you've backed up your computer, you can turn your attention to all the photos, records, films, and cassettes that you'd like to integrate into your digital collection and figure out which task to tackle first. It might be perfectly clear to you which items are most important, in which case you can quickly fill out the Priority Score column of your Media Collection Inventory (page 11). But if you're not sure where to start, you can use the following scoring system to help you set your priorities. You'll give each item on your list a score from 1 to 20, and record that number in the Priority Score column. The highest scores will come first, so you'll start with the items that get a 20 and work your way down to 1.

The Top Five Priority Scores

20. Give any item that you need for an upcoming event an automatic 20. This might be a video to show, a slideshow of photographs, or music.

19. Check your analog media for decay—and fading in the case of photos—and give anything that looks like it's deteriorating a 19.

18. If you have film, analog video tapes, CDs or photos that you've decided to have digitized by a service instead of doing it yourself, give those items an automatic 18 and get them out the door. The process is quick and easy enough that there's no reason to put it off. You can read more about scanning, transfer, and ripping services in the chapters on photos, videos, and music.

17. Give media that require rare devices for playback a 17. As the devices become harder to find, it may become more expensive to purchase and repair them.

16. Take a look at the remaining items on your Media Collection Inventory and note the ones that could not be replaced if they were lost or damaged. Now ask yourself which one you would most regret losing, and make it number 16.

Once you've assigned those top priorities, use the list on page 14 to score everything else. Tally the number of Yes answers you gave for each item in your Media Collection Inventory, and record that total in the Priority Score column. If you end up with more than one item with the same score, ask yourself which you'd choose if your house were on fire and you could save only one of them, and then score the items according to your immediate response.

Before you start digitizing any analog media, organize the digital files of the same type that are already on your computer so that you have an organized structure to integrate your new digital files into. For example, before you start scanning any photos, put your existing digital photos in order. You can find details on how to organize your photos, music, and videos in the chapters on those topics.

Archiving on Discs

Archiving your media on optical discs can be an excellent way to preserve your data by using technology that is affordable and widely accessible. Although you might use hard drive storage to regularly back up current folders and files that you frequently add to or change, discs are useful for backing up sections of media collections that won't change. For example, you might archive your photographs from previous years on disc. Discs can also be useful for backing up media with a standalone disc burner when you're away from your computer or simply for convenience.

If you have a lot of disorganized files on your computer, it's best to put them in order before archiving them to disc. Take a look at the sections of this book on creating a system to organize your photos, videos, and music before you start burning discs.

There has been some controversy over the longevity of optical discs, and various studies have reached different conclusions. But most reliable sources indicate that high-quality optical discs can last for decades if properly burned and stored—which raises the question of how to tell whether the discs you're using to archive your media are any good. Disc quality varies widely, and as yet, there is no industry standard for rating disc quality on product packaging. However, an international standard for evaluating DVD quality, ISO 10995, does exist. As the industry adopts use of that standard for quality testing, its name may begin to appear on DVD package labeling to indicate discs that meet the international standard for archival quality.

Without reference to a standard, terms such as "archival" can mean whatever a manufacturer decides is appropriate. That said, it's likely—although not guaranteed—that discs labeled "archival" are made of better materials and receive better quality control than other discs from the same manufacturer. For archiving purposes, never use generic discs or those sold under office supply store brands. Buy from a reputable name brand and look for discs that have special coatings to protect against scratches and fingerprints.

Optical Disc Types

There are numerous types of optical discs. The ones you are able to burn files to will depend on the capabilities of your optical disc drive. If you don't know which formats your drive supports, and whether it can write to double-layer discs, go to the support section of Nero's Web site (www.nero.com) and download the free InfoTool. This little program will tell you more than you probably ever wanted to know about your optical drive. The following widely supported disc types are the main options you should consider using to archive your media:

- **CD-R.** These discs offer capacities of 700MB each.
- **DVD-R and DVD+R.** These discs typically offer capacities of 4.7GB or 8.5GB for double-layer discs.
- **BD-R.** Recordable Blu-ray discs offer capacities of 25GB or 50GB in double-sided discs. This is a newer technology, and optical drives that write to Blu-ray discs tend to be more expensive. However, if you're archiving large amounts of media, you may find the investment worthwhile.

Ten Steps to Preserving and Organizing Your Media on Discs

✔ **1. Check your optical disc drive.** Make sure your drive firmware is up to date. Go to the support section of the drive manufacturer's Web site and download any firmware update available. (Nero's InfoTool can tell you which version your drive currently has installed.) The quality of your drive will affect the quality of your disc burning, so if you're buying a new drive, look for product reviews online first. If you have a drive that's more than a few years old and you get poor results when you burn a couple of test discs, consider replacing it. You can install a new optical drive in your desktop computer the same way you install an internal hard drive (see page 22), except that you mount it in the bay out of which you took your old optical drive. Or you can use an external optical drive that connects via USB.

Information sources ▸▸
These online sources provide product reviews, technical information, and forums.

Blu-ray.com www.blu-ray.com
Blu-ray Disc Association www.blu-raydisc.com
CDFreaks www.cdfreaks.com
CDRLabs www.cdrlabs.com
DVD Forum www.dvdforum.org
Optical Storage Technology Association www.osta.org

✔ **2. Burn different media types to separate discs.** Archive your photos, music, videos, and documents on separate sets of discs unless you have a good reason not to (for example, if you want all of the files related to a particular event on one disc). Mixing media types randomly will lead to disorganization and make it difficult for you to find archived files. Once you've created systems for organizing your photos, music, and videos, as described in the chapters on those topics, you can simply copy the files and folders in each collection to disc in the same order in which they're organized on your hard drives. This approach will also help you create clearly organized labels for your discs.

✔ **3. Verify your success with each disc after burning it.** Use a software program to burn your discs that offers a verification function, and run the verification after you burn each disc. The verification process checks the data on your disc to make sure that the disc was recorded correctly and to let you know if there's a problem.

✔ **4. Use a slow burn.** Some discs and drives are rated for burning at very fast speeds, but the fastest speed is usually not optimal. Slower write speeds are generally less prone to creating disc errors. Some drives offer functions that can evaluate the characteristics of your disc before burning it to determine the optimal speed.

✔ **5. Don't overwrite archive discs.** Even if you use rewritable CD-RW or DVD-RW discs to archive your media, it's best not to overwrite them multiple times, to avoid increasing the chance of introducing errors.

✔ **6. Label your discs.** Good options for labeling include a pen that is made specifically for disc labeling; a LightScribe or Labelflash system, if your disc drive offers that feature; or a specialty disc-labeling product that uses archival materials. Don't, however, use adhesive labels or write on discs with regular pens. You can purchase disc-labeling pens from the archival supply sources listed on page 184. If you have a printer that can print labels directly on discs, make sure you buy discs that are compatible with the type of label printing technology your printer uses—inkjet, for example. Devise a disc labeling system that makes sense to you and is easily understandable to anyone else who might use the archive in the future. Note the date when your disc was burned so that you know when it's time to check its data integrity. If you use double-sided discs that can't be written on, keep them in their jewel cases and create case labels.

0 Zero Minute Disc Archiving

If you'd like to archive your photos, music, and personal documents on disc, but you don't have time to spend clicking through software options and selecting files to burn, there may be a completely automatic disc archiving system that's right for you. Storage Appliance Corporation's Clickfree and HP's SimpleSave provide discs that automate the process of archiving specific kinds of media files when inserted in a disc drive. Clickfree and SimpleSave discs are available in a version that saves photos. Clickfree discs also come in a Music version and an Office version that saves several types of e-mail and document files. The Photo and Music versions find and archive all image and audio files on your system.

All you have to do is put the disc in the drive, hit Go when the disc's built-in software opens, and take it out when the burning process is complete. (You can also choose an options menu first, if you want to select specific file types to be burned.) If more than one disc is required to hold all of the photo, music, or document files on your computer, the Clickfree or HP software tells you how many discs you'll need before it starts burning files, then requests that you insert another disc once the previous one is full. The folders your original files are stored in are replicated so that your files are organized on the disc. When you insert a recorded Clickfree or SimpleSave disc in a drive, the software opens and asks if you'd like to restore the files to your computer or play them.

Disc Resources

Disc Manufacturers
There are many companies that sell optical discs. These are a few that are widely considered to produce some of the better products.

Maxell www.maxell.com
Mitsui MAM-A www.mam-a.com
Taiyo Yuden www.t-yuden.com
Verbatim www.verbatim.com

Disc-Burning Software Makers
These companies make advanced but affordable disc-burning software.

Nero www.nero.com
Padus DiscJuggler www.padus.com
Roxio www.roxio.com

Disc and Drive Utility Software
These free or inexpensive programs can show you information about your optical drive and inserted discs, run surface-scan and read tests to tell you if your discs contain errors, and perform numerous other detailed tests for the technically inclined.

DVDInfoPro www.dvdinfopro.com
Nero DiscSpeed www.cdspeed2000.com
Plextor PlexTools www.plextools.com

✔ **7. Keep your discs clean.** Hold them by the edges and avoid touching the surfaces. If your discs get dusty, dirty, or smudged, clean them with a soft cloth and rubbing alcohol or a special optical disc-cleaning solution. Wipe them off from the center to the edge, not in a circular motion. Make sure the tray of your disc drive stays clean, too.

✔ **8. Store your discs properly.** It is of the utmost importance to keep your archive discs away from sunlight and other sources of ultraviolet light. High temperatures, wide temperature fluctuations, humidity, and anything scratchy can also cause damage. If you live in a very humid area, purchase small desiccant packs from an archival supplies source (page 184) and put them in the jewel cases with your discs. Storing your archive discs vertically in individual jewel cases to prevent them from getting bent is the best approach. You shouldn't use your archive discs frequently. If you need to have your files on disc for frequent use, burn and store a separate set of archive discs.

✔ **9. Check your disc archive once a year.** At the end of the year, or whenever your digital housekeeping date of choice occurs, give your disc archive a once-over and make sure everything is in order. Use a tool such as Nero DiscSpeed, PlexTools, DVD InfoPro, or ISOBuster, to run older discs through a surface scan or do a read check. If the software discovers errors, replace the disc with a new one. You can use the disc duplication features of advanced disc-burning software to copy the contents to a new disc.

✔ **10. Create a dual-disc archive.** To do your utmost to ensure that your media files are preserved, you can create a second copy of your archive on a set of discs made by a different manufacturer than the discs used for the first set. Any manufacturing flaws that might affect the lifespan of the first set will not be replicated in the second set.

Backing Up and Synchronizing

The Goal:

Set up a local backup of your media and other files, or synchronize files between two or more devices.

What You Need:

Backup or sync tools included in your operating system or additional software.

Time Required:

Ten minutes for a basic setup; more time to set custom options.

If there's one thing that stops people from preserving their digital media the way they should, it's the onerous prospect of backing up their data. Backing up just sounds like a pain. But in fact, with current software options, setting up a basic backup or sync is one of the quickest and easiest things you can do on a computer. And if you set your backup to run automatically on an ongoing basis, you won't ever have to touch the software after the initial setup.

Even if you spend a little extra time customizing your backup, you'll be hard pressed to stretch the task out to more than half an hour. Many current backup and sync programs offer very clear and simple interfaces and guide you through the basic setup with straightforward wizards. The basic steps are to select a drive or computer to back up and choose a drive or device to back up to, or to select two drives or devices to sync with each other. The process can be as simple as that. If you try a backup program and find it overly complicated, look around for a replacement that you're more comfortable with.

You should expect the backup software to take a substantial amount of time to complete an initial full backup after you set it up. However, you won't have to do anything while it runs. The amount of time a first backup takes depends on how much data you're backing up and the type of connection between the source and the device where the backup will be stored. Your initial backup might take hours, or even days. But subsequent backups should be much faster, since only new and changed items will need to be copied to your backup storage device.

Backup and Sync Approaches

Some software programs take just one of these approaches, but many provide a combination of them. Depending on your needs, you might use one approach or more.

Drive imaging. This approach makes a copy, or image, of an entire computer or drive, including all of the installed software and settings, in addition to files and folders. This is the most comprehensive backup method, but it also requires the most storage space, and the software that created the image must be used to restore a system on a new machine after a computer or drive failure.

File backup. This approach backs up files and folders, but does not save installed software or settings.

Synchronization. Sync tools keep the contents of selected drives or devices in sync, so that their contents are identical. That means that when you make a change to a file on one device, its twin on the other device is updated. Synchronization can be useful for easily making new media files available to multiple computers and devices, especially with collections of files that you will add to but won't edit, such as commercial music collections. On the other hand, you may not want to sync files that you will edit and create multiple versions of—such as photo files—since most sync tools don't provide versioning.

You can use sync software to keep computers in different locations in sync by using a small external hard drive or a USB flash drive as an intermediary—sync one computer to the drive, then sync the drive to the other computer. This system can work as an alternative to using an online backup or syncing computers in different locations via the Internet, and may be more affordable if you have a large amount of data.

Backup Software Sources

There are several places you can find backup tools and software:

Your operating system. If you have Windows XP, Windows Vista, or a current Mac operating system, you already have built-in backup tools. They don't offer all of the features that most dedicated backup programs do, but they will be sufficient for many people, and they're free.

In the box with your storage device. Many external hard drives and network-attached storage devices come with backup software.

Integrated into your media or photo management software. Some advanced media management software includes backup tools, usually for a specific kind of files, such as photos.

Ten Backup and Sync Features to Look For

✔ **1. Continuous backup or sync.** A backup or sync program that can be set to back up files whenever they are changed or added ensures that the latest versions of your files are secured. If the backup destination or synced device is disconnected, the update takes place as soon as it is reconnected.

✔ **2. Versioning.** This feature saves multiple versions of your files so that you can retrieve them from different points in time, without recent changes. Some software lets you select the number of versions to retain so that you can balance security with storage space. This feature is common in backup software but usually unavailable in sync software.

✔ **3. Flexible criteria.** Some software lets you choose to back up or sync files in particular categories, such as music or photo files; many programs let you select specific file types to back up. This can be a useful feature if you want to conserve storage space. Exclusions are another selective backup tool. Being able to exclude specific file types from a general backup allows you to preserve all of your irreplaceable personal media and documents without taking up storage space with large commercial media files such as movies and TV shows. Some software allows you to exclude files that exceed a size or date limit.

✔ **4. Scheduling.** All but the most rudimentary backup and sync tools allow you to schedule updates to take place on a regular basis. Even if you use a program that offers continuous backup of files, you might want to schedule full drive imaging backups to make sure that you are preserving your latest software and settings.

✔ **5. Simple and flexible restore tools.** Look for software that makes restoring your files or system easy and lets you choose between restoring individual files and folders or your whole system. To allow you to recover from a total system failure or main hard drive crash, some programs can create a bootable disc that you can use to restart your system. This is called a "bare metal" restore.

✔ **6. Reporting.** Backup software should create a log that records the backup process and lists any files that could not be backed up. Some software can be set to e-mail you a log file or a confirmation that a backup has taken place.

Backup Resources ⏩

Backup and Sync Software Makers
If your operating system or your media or photo management software doesn't provide all of the backup or sync features you want, try a standalone backup program from one of these software companies.

Acronis www.acronis.com
Centered Systems www.centered.com
EMC Insignia www.emcinsignia.com
Genie-Soft www.genie-soft.com
Host Interface International www.hostinterface.com
Macrium www.macrium.com
Memeo www.memeo.com
Nero www.nero.com
NewTech Infosystems (NTI) www.ntius.com
NovaStor www.novastor.com
Novosoft www.handybackup.net
Paragon Software Group www.paragon-software.com
Prosoft Engineering www.prosofteng.com
Roxio www.roxio.com
Siber Systems www.goodsync.com
Softland www.backup4all.com
StorageCraft Technology www.storagecraft.com
Symantec www.symantec.com
TGRMN Software www.tgrmn.com
Titan www.titanbackup.com

✔ **7. File compression.** Some software can automatically compress backup files with a standard format such as ZIP or with a proprietary format. This will help you conserve storage space. However, if the software uses a proprietary compression format, you need to use the same program that compressed the files to open them again.

✔ **8. Backup and sync destination options.** Some programs let you back up to or sync with a wider variety of destinations than others. Possible destinations include external hard drives, networked computers and drives, USB flash drives, online storage and FTP sites, and optical discs. If you want to back up to optical discs, choose a program that offers disc spanning so that you can back up large amounts of data to multiple discs in a continuous process.

✔ **9. Backup search tools.** Some software allows you to search for specific files in your backup so that you can restore files selectively.

✔ **10. Backup of files in use.** Some software can back up files while they are in use by a program on your computer.

Backing Up with Your Operating System's Tools

Windows Vista. To set up automatic backups in Windows Vista, go to the Control Panel and click on the Backup and Restore Center. Schedule a backup to run daily. You can also manually run a complete system backup that preserves software and settings in addition to files. Files or an entire system can be restored from the Backup and Restore Center. If you're unable to start Vista after a system crash, you can insert your Vista OS disc to start the restore process. Windows Vista can also automatically preserve previous versions of files. To make sure this feature is enabled, go to Control Panel/System/System Protection. Under "Automatic restore points," make sure the boxes next to the listed hard drives are checked. To retrieve a previous version of a file or folder, right-click on it in Windows Explorer and select Restore Previous Versions. (Note that Windows Vista Home Basic does not offer all of the backup features included in the other editions of Vista.)

Windows XP. XP users running the Professional edition can find the Windows Backup Utility by clicking on the Start menu and going to Accessories/System tools/Backup. XP Home users will likely need to install the Backup Utility from the XP installation disc, since it is not installed automatically with the operating system. To do so, insert the disc and wait for the disc options screen to appear. Then select Perform Additional Tasks/Browse This CD/ValueAdd/MSFT/NTBACKUP. Install the ntbackup.msi file. Once installed, the Backup Utility will give you tools for scheduling automatic file backups, as well as running complete system backups that preserve software and settings in addition to files. Files or an entire system can be restored from the Backup Utility. It can also create a bootable disc.

Apple Mac OS X Leopard. Backing up on a Mac with a current operating system is very simple. Go to the Time Machine application and select a backup destination. It will automatically back up your whole system, with versioning and flexible restore options. Customize the backup settings as you wish.

Keeping Your Media in the Cloud

When technophiles talk about online storage, they often refer to their data being "in the cloud," as though there were a magical, Oz-like domain in the sky where all the stuff that's accessible via the Internet resides. In reality, when you use an online service to store your digital media and other files, they reside somewhere a bit more prosaic: a data center. Ideally, the files are stored on servers in more than one data center so that they're protected from any disaster that might occur at a single location. Regardless of the way you prefer to envision the place where your files reside, online services can provide an excellent way to back up and share your files, and to make them available on the different devices you own.

These are the main types of online services available:

- **Backup.** A backup service allows you to upload your files to a remote server, so that they can be restored in the event that you lose the files you have stored at home.
- **Synchronization.** Sync services let you designate files and folders to be synchronized with an online server. That means that current versions of your files will be available to you on a password-protected private Web site from any Internet-connected computer, and sometimes from mobile devices such as cell phones.
- **Remote access.** A remote-access arrangement doesn't store your files online, but gives you Web access to your computers, network-attached storage device, or home server, as long as it's turned on and connected to the Internet.
- **Peer-to-peer storage.** The relatively new peer-to-peer type of service lets you back up your files by storing them on other people's computers instead of in a data center. It encrypts your files so that no one can open them, then distributes multiple copies of them to free space on the Internet-connected computers of other people who use the service. One of the main advantages of peer-to-peer storage is that it is very inexpensive in comparison to purchasing space on a server, allowing you to store large amounts of data for an affordable price. To use a peer-to-peer system, you generally have to leave some space free on your own computer so that other users' encrypted files can be stored there. In effect, this means you are trading your hard drive storage for online storage, which is a good deal because hard drive storage has a much lower cost per gigabyte.

Costs and Requirements

Online backup and sync services often offer some amount of storage for free, then charge for additional space on a monthly or annual subscription basis. Some charge a monthly fee per gigabyte of space used and amount of data transferred, while others offer unlimited storage at a fixed annual price. Many services have limits on the size of individual files that you upload, so check to make sure the one you select will be able to handle all of the files you want to keep in the cloud. Most offer some type of free trial so that you can see if a service works well for you before buying in and uploading a lot of data.

There are numerous services that allow you to back up unlimited amounts of data for an annual fee in the range of 50 dollars. But in that price range, the features available are usually limited to simple backup and restore options. Other providers, such as Windows Live and Apple MobileMe, offer a whole suite of online tools and services. Expect to pay higher fees for more feature-rich service packages. You may want to combine online backup or sync of a selection of files with remote access, a cost-effective way to secure important files while preserving access to your whole media collection from locations away from home. Another option to consider is using an online image gallery service (see page 108) to back up photo and even video files online, while storing or synchronizing other types of files through a general online backup or sync service. Many online photo gallery sites offer more competitive rates on online storage, since they can make money through means other than providing storage space, such as selling prints.

In order for an online backup, sync, or remote access system to work well for you, you need to have a fast Internet connection. Trying to use a dial-up connection or even slow DSL is like trying to get your data to the cloud by taking the stairs. If you can't get a fast DSL, cable, or other high-speed broadband connection, you may want to consider the alternative of offsite storage, such as a hard drive or optical discs kept in a safe deposit box, for important files.

How Online Backup and Sync Services Work

To use most online backup and sync services, you typically download the service's software to each computer where you have files that you want to keep in the cloud. Then you designate folders or whole drives to be backed up or synchronized. Some services provide access to your online storage or remotely accessible computers through their own software or a Web browser, while others integrate their features into your operating system's file browser so that your online storage area appears as a local drive. For example, if you use Jungle Disk, when you open Windows Explorer or Mac Finder, you will see a J drive for Jungledisk, along with your local C drive. You can click on the drive letter to access files and drag files and folders to the online "drive" to archive them online, just as

Backup and Sync Services ▸▸
Apple MobileMe
www.apple.com/mobileme
4Shared www.4shared.com
backup.com backup.com
Bluestring www.bluestring.com
Box.net box.net
Carbonite www.carbonite.com
DropBoks www.dropboks.com
Fabrik www.fabrik.com
File Den www.fileden.com
FlipDrive www.flipdrive.com
IBackup www.ibackup.com
IDrive www.idrive.com
Intronis eSureIT Home
www.intronis.com
Iomega iStorage
www.iomega.com
Iron Mountain
backup.ironmountain.com
More on page 180-181

you would do with a local drive. If you use a Windows Home Server (see page 22), you can select a service (such as Jungledisk) that integrates online backups into its features.

If you use an online storage service that provides FTP access to your storage space in the cloud, you can set your local backup software to automatically back your files up online in addition to performing backups to hard drives and other local storage. FTP (File Transfer Protocol) is simply an efficient system for transferring files to and from an Internet-accessible server. Many of the advanced programs from companies listed on page 38 can perform FTP backups. The typical process of backing up to an FTP server is quite simple: You enter your FTP user name and password, then set the software to automatically back up to the FTP server at regular intervals.

If you decide to keep a large amount of data in the cloud, expect your initial online backup or sync process to take a long time. Uploading more than ten gigabytes or so might take days or even a week, depending on your Internet connection speed. You will still be able to use your computer during the process; just remember to leave it turned on until the initial upload has been completed. Successive backups and sync processes should be much faster, since the service will only need to upload files that have been changed or added to keep your online storage current.

Ten Online Storage and Sharing Features to Look For

✔ **1. Partial or block-level backups.** This type of backup can update parts of files instead of replacing whole files with new versions. This helps it quickly update your backed up or synchronized files in the cloud as you make changes to them on your computer, and makes more efficient use of your Internet connection.

✔ **2. Mobile access.** If you have an advanced mobile device such as a smartphone look for a service that lets you view the files you keep in the cloud on the phone and upload files directly from it. Some services give you mobile access through an Internet browser, while others may require you to download and install a mobile application. You must have a data service plan for your mobile device in order to use mobile access.

✔ **3. Image galleries and media players.** Some online services let you arrange the image files you upload into Web albums, just like the albums on a dedicated online gallery site. Some services also include Web-based media players that allow you to play music and videos that are stored on a remotely accessible computer or kept in the cloud, without downloading them. Such a remote-access feature can let you listen to or display your whole media collection from any Internet-connected computer or compatible mobile device, wherever you are.

✔ **4. Web site and blog hosting.** In addition to offering online backup and sync services, some feature-rich providers give subscribers simple tools for creating a Web site or blog, as well as space to store it online.

✔ **5. File sharing and export tools.** Some services provide ways for you to share the files that you keep in the cloud with other individuals or groups, usually by giving them direct access to a shared folder or gallery on the Web. Export tools can give you another way of distributing your media to others, by publishing them directly from your storage service to an online destination such as a social networking or gallery site.

✔ **6. Encryption.** All services that give you Web access to your files allow you to protect the access with a password. Many services also encrypt your files as they're uploaded and while they're stored on a server in the cloud, so that no one but you and the people you share them with can open them.

✔ **7. Flexible uploading.** Look for a service with software that can pause and resume uploads when it is backing up or synchronizing your files. It should be able to pick up where it left off if you shut your computer down while files are being updated, and it should be able to optimize its use of your Internet connection's bandwidth so that it doesn't affect you if you're using a Web browser while a backup or sync is taking place. When you're evaluating a service during a trial period, try surfing the Web while it's uploading, and make sure your connection is as smooth as usual.

✔ **8. Web access to stored files.** Some online backup services simply allow you to store copies of your files in the cloud, and to restore them to your computer in the event that they're lost. Others also allow you to view the files you have stored online through a Web browser, without having to download them. If you travel or use different computers in multiple locations, a service that gives you Web access is probably the best option.

✔ **9. Versioning.** Although it's not a common feature, some services that offer online backup provide versioning tools. This feature allows you to save and restore multiple versions of files as they are changed over time.

✔ **10. Web-based applications.** Some services, such as Glide, include Web-based applications in the set of tools they provide subscribers. Web-based applications can include software for working on documents in addition to media players, so that you can not only access, view, and play your files with a Web browser, but you can also work on them, without installing any software on a local computer.

Protecting Your Media from Disaster

You can and should protect your computer, external hard drives, and other digital media-storage devices from power surges by plugging them into a surge protector, instead of directly into an electrical outlet.

If you live in an area prone to floods, fires, earthquakes, or other natural disasters—or on a boat, in a desert, or any place where your media are likely to be exposed to water, dust, and other environmental dangers—you can protect your media collection and personal documents by keeping the most important and irreplaceable portions in a protective container or device. You might also choose to store your most treasured photographic prints and important paper documents this way.

Consider archiving important digital media and documents online, too, if you're concerned about disasters or theft. Prices for disaster-proof storage start at less than $100.

Protective storage devices are available in a variety of fireproof, waterproof, and impact-proof models, including the following:

- **Home safe.** A strong, secure box with a combination, keypad, or biometric (i.e., fingerprint-reading) lock. Some safes can be bolted to a floor or wall to protect against theft.
- **Security chest or box.** A box that may not be as secure as a safe for protection against theft, but offers protection from natural disasters.
- **Optical disc file safe.** A safe or a security box that's designed specifically to store optical discs.
- **Laptop safe or case.** A container whose dimensions accommodate a laptop computer.
- **Document file safe or chest.** A safe or box with an interior designed for storing paper documents and photographic prints.
- **Waterproof case.** A case that will protect your media from water, dust, and impact damage, but isn't fireproof.
- **External hard drive.** SentrySafe makes external USB hard drives that meet UL (Underwriters Laboratories) standards for being fireproof and waterproof.

What to Look for in a Disaster-Proof Storage Container

Fireproofing. To protect digital media during a fire, fireproof containers maintain a temperature lower than 125 degrees Fahrenheit and a humidity level below 80 percent. Fireproof safes, boxes, and drives are rated to withstand fire for specified periods, ranging from half an hour to three hours.

Waterproofing. You can purchase safes and secure boxes that offer protection from water and humidity in addition to fireproofing and strong impact resistance. If water is your main concern, there are also cases that offer protection only from water, dust, and low impacts, and are often more portable.

Impact proofing. Protective containers and drives can undergo standard drop testing to ensure that they will not break open and will protect their contents in the event of a strong impact from a fall or from being struck by another object.

Independent certification. Make sure that the storage option you choose has been tested and certified by Underwriters Laboratories—or another independent organization such as Intertek—to meet specific fireproofing, waterproofing, and impact-proofing standards. UL and Intertek are independent product safety certification organizations that create safety standards and test products.

Digital-media-specific storage. Requirements for protecting digital media such as optical discs are different from those for paper and other materials. Look for a container that meets those requirements for protection from fire, water, and humidity, and that allows optical discs to be stored vertically. Some larger safes have separate compartments for digital media and papers.

Whichever choice you make, try to check its integrity and viability each year during the recommended time you edit and organize your photographs, an exercise mentioned in the next chapter.

Protection Resources ▸▸

Independent Product Testers
Intertek www.intertek-etlsemko.com
Underwriters Laboratories www.ul.com

Safe Manufacturers
Cobalt Safes www.cobaltsafes.com
Fire Fyter www.firefyter.com
FireKing Security Group www.fireking.com
Honeywell Safes www.honeywellsafes.com
SentrySafe www.sentrysafe.com

Waterproof Case Makers
OtterBox www.otterbox.com
Pelican www.pelican.com
Seahorse www.seahorse.net
Underwater Kinetics www.uwkinetics.com

Recovering Lost Digital Media

It's bound to happen sooner or later: you will lose digital media or document files. It might happen because an optical disc gets scratched, because your hard drive started making funny clicking noises and then just gave up altogether, or simply because you goofed and deleted a bunch of files accidentally.

When this practically inevitable event occurs, don't panic if you don't have a backup of the lost files. There are steps you can take to recover the data you've lost. They won't work in every case, but there's a good chance you'll be able to get at least some of your files back in many if not most cases. Usually when you can't retrieve data from a hard drive, optical disc, or flash-memory drive or card, the files are still stored on the device. The problem is that the software structure or physical mechanism that allows you to get at them has been damaged.

Here's what to do:

✔ **1. Stop using the drive, card, or disc.** If your hard drive seems to be suffering from a mechanical failure—clicking and chugging noises, or no apparent drive activity at all, will tip you off—trying to use it can exacerbate the damage. When the problem isn't physical, adding new software or files might overwrite the files you're trying to recover.

✔ **2. Don't reformat.** If you reformat a hard drive or flash memory card or drive, it will be impossible to recover your files without using sophisticated recovery tools or services. The risk of files being irretrievable will also increase.

✔ **3. Try a different optical drive or card reader.** If an optical disc is unreadable, try using a different disc drive. A newer drive may offer better error correction and be able to read the disc. If that works, it's probably time to replace your old optical disc drive. You can install one in your desktop computer by following the same process you would use for an internal hard drive (see page 22), or you can buy an external optical drive that plugs into your computer via USB. For memory cards, try inserting the card in a different reader or device to make sure that the problem is the card and not the reader.

✔ **4. Run data recovery software.** If you're not dealing with a mechanical problem or physical damage, choose software designed to recover files from the specific storage device you're dealing with—a hard drive, optical disc, or memory card. If you are trying to recover photo, music, or video files, look for software made specifically for that purpose.

When using hard drive recovery software, you will need to install and run it on a hard drive other than the one holding the irretrievable data so that you don't overwrite any of the files you are hoping to retrieve. If you're trying to recover from a computer with a single hard drive, the easiest option is to connect an external hard drive to it and install the recovery software there.

Using recovery software with optical discs, memory cards, and USB flash drives is more straightforward, since you can install the software on your computer as usual, and then run its recovery process on the inserted or attached disc, card, or flash drive. Look for data recovery software that is available in a demo version that shows you a list of the files it can recover when you run it on your problem drive or disc. That way, you can see whether the software is able to detect your lost files before buying it.

✔ **5. Try a disc-resurfacing product or service.** If you have cleaned your CD or other optical disc (see page 35) but still can't retrieve the data on it, check it for scratches. Even fine scratches can throw off optical drives and prevent data from being read correctly. You can purchase a device that polishes the surface of the disc in order to smooth out scratches and make the disc readable again. Services that use professional machines to polish discs are also available online and from some stores that sell CDs and DVDs.

✔ **6. Send the drive, disc, or card to a recovery service.** If you can't retrieve files from your hard drive because of a mechanical problem, a recovery service is your only option. Even if you're not dealing with a physically damaged hard drive, you may benefit from the knowledge, experience, and tools of trained technicians. Choose a service that won't charge you if it can't recover your files. Hard drives should be packed in anti-static bags and thick foam padding or bubble wrap before shipping.

An Ounce of Prevention

Diagnostic software can keep tabs on your hard drive and let you know if it's at risk of failing. It won't catch every glitch, but it will pick up on many problems so that you have time to copy your data to a new drive before a crash.

Look for software that includes error checking, surface scanning, and S.M.A.R.T. (Self-Monitoring, Analysis and Reporting Technology) tools. You should be able to set up the software to run automatically on a regular basis. You can also manually run the Check Disk on Windows systems to detect certain types of disk errors.

Recovery Resources ››

Memory Card Recovery Software
This is a selection of the programs that can recover files from the flash memory cards that are used in digital cameras and other portable devices. Some memory cards also come with recovery software.

Best IT Solutions PhotoOne Recovery
www.photoone.net

DataRescue
www.datarescue.com
Disk Doctors Digital Media Recovery
www.diskdoctors.net
Galaxy Digital Photo Recovery
www.photosrecovery.com
ImageRecall Don't Panic
www.imagerecall.com
More resources listed on page 181

77mm MC UV(0)
24-105mm 1:4
MACRO
0.7
ft
m
24-105mm
24
35
50
70
105
AF·WB
DRIVE·ISO

CHAPTER II

Photographs

Introduction:

For most of us, few of our personal possessions are closer to our hearts than our photographs. And while music, videos, and even personal documents can often be replaced, our albums, shoeboxes, and computers may be full of images that aren't preserved elsewhere, and can never be captured again if they are lost.

Digital photography makes it possible to take more pictures than ever before, do more creative things with them, and share them with distant friends and family in new ways. But it also brings new dangers to their preservation, and makes it easy for us to end up with a computer full of unmanageable, unidentifiable, and unused image files very quickly.

In this chapter, you'll find fast and easy ways to get your digital photo collection organized and to keep it safe, to digitize and restore your old film photos, and to take advantage of new digital tools for displaying, sharing, and preserving all of your photos.

Digitizing Prints, Negatives, and Slides

The Goal:

Scan your collection of prints, slides, and film and save the images in a digital format, letting you add the photos to your digital image collection, share them online and on disc, and archive them for safekeeping.

What You Need:

Film or flatbed scanner; cleaning solutions, brushes, and cloths; music to listen to while you're scanning.

Time Required:

The time can vary widely, depending on your scanner and software, the settings you use, and the size and number of photos. To make a rough estimate of the total scanning time required to digitize your collection, allow one minute for each image you're scanning.

Just about everyone who is old enough to have finished high school has at least a shoebox full of old prints somewhere. Many of us have lots of boxes and albums filled with 4x6 prints, envelopes of long-forgotten negatives, and a shelf somewhere in the back of a closet that's packed with carousels of old slides.

Instead of letting these photographs sit on the shelf, you can include them in your collection of digital photos by scanning them and saving them as digital image files. This also allows you to add them to your online gallery so that friends and family can view them and order prints easily. Once you have digital versions of the photos, you'll also be able to do creative things with them in image editing software and make new prints of them yourself. If you have a lot of faded, damaged prints, but were wise enough to save the negatives, you might be surprised at how great the pictures look when you scan the film and make new prints of the digital files. Scanning your photos will also give you a way to archive them online or in a remote location that will be safe in case of a house fire, a burglary, or some other catastrophe.

But there's a downside to digitizing your analog photo collection, too: It's a project. No matter how efficiently you go about it, scanning all that stuff takes time, and frankly, the process is not a barrel of laughs. Scanning photos is a repetitive task that requires attention to detail and a grasp of scanning basics to produce worthwhile results. If you're not very comfortable with new technology and software, hate repetitive tasks, or simply have other fish to fry, I strongly recommend that you consider having your photo collection digitized by a scanning service. Be honest with yourself about whether you will actually sit down and get through all the scanning, after the initial thrill of playing around with a new scanner fades. You can read more about scanning services on page 68.

If you're still reading this page, then you must be up for a project. Scanning options can get very complicated, and there are many advanced tools available to serious photographers and graphic artists. The goal here is to give you just the information you need to choose the right scanner, go about the scanning process in the most efficient way, and create a high-quality digital archive of your analog photos.

Choosing a Scanner

If you already own a scanner that's more than a couple years old and have a lot of photos to scan, consider buying a new one. Newer scanners will do the job faster and give you greater quality and more tools for your money than the models available just a few years ago.

The first thing to determine before you choose a scanner for digitizing your collection is whether you will be scanning film, prints, slides, or a combination of those media. Also consider whether you have any film in formats other than 35mm. The 35mm format has been by far the most common film size used in cameras over the past forty years or so, but you might have smaller APS or larger medium-format film in your collection as well. (Note, by the way, that there are no scanners that can handle undeveloped film. If you have rolls or cartridges of undeveloped film, take them to a store or photo lab to have them developed, and then scan them.)

If you saved the negatives of any photos you printed from film, by all means, scan the negatives instead of the prints. Film retains more image detail than prints, is often better preserved, and can be touched up automatically with infrared scanning technology.

There are only a few reasons why it might be a good idea to scan prints of your images instead of film, when you have a choice. Scan the prints rather than the film:

- If the film is heavily damaged
- If you don't have access to a scanner with reasonably good film scanning capabilities
- If the prints were made by hand or with a special process, and you want the digital image to reflect the artistic choices that the photographer or printer made when creating the prints. (In this case, you might want to scan both the prints and the film.)

Once you determine what type of media you will be scanning, you can select the most appropriate tools that your budget will allow from the following options:

Flatbed scanner with no film or transparency adapter If you're scanning prints only, this less expensive type of scanner will do the job. If you're on a budget and are purchasing a scanner in the $100 range, consider getting a multifunction or all-in-one device. These include other functions such as printing, copying, and faxing, and the scanners in inexpensive models are generally just as good as standalone scanners that cost roughly the same price. You'll get more for your money with no sacrifice in scan quality.

Flatbed scanner with a film adapter
The most economical way to digitize a collection that includes film, slides, and prints is to buy a flatbed scanner that has a built-in transparency adapter and film scanning functions. Scanners with the essential features needed to capture good-quality scans start at about $200. Current flatbed models that cost about $250 or more can produce film scans of very good quality.

Film scanner
As the name suggests, a film scanner digitizes film only and does not handle prints. Prices for dedicated film scanners with the features necessary for high-quality scans start at about $300 and go up to about $2000. In general, a high-quality film scanner can produce better film scans than a typical flatbed with a film adapter. If you will be scanning film in a format other than 35mm, expect to pay a higher price for a scanner that handles multiple formats. If you can't afford one, a flatbed scanner with a multiple-format film adapter is a more affordable option.

Any film scanner worth its salt can scan both film strips and slides, although some models may require you to purchase a slide adapter separately, increasing the cost. Some film scanners are compatible with autofeeders that will accept a whole stack of slides at once, as well as attachments that can automatically feed an entire roll of uncut 35mm film into the scanner. These accessories cost extra. Inserting film strips and slides into an autofeeder makes the process of scanning large quantities of film or slides go faster.

Slide scanner
Some scanners are made specifically for scanning large collections of slides. They have automatic feeders that can handle stacks of about fifty slides at a time. These are generally high-quality scanners that cost more than $1000.

Essential Scanner Features

Scanners offer a wide range of features and functions. Two are essential for producing high-quality scans in a reasonable amount of time:

Batch scanning

Batch scanning allows you to set up the scanner to automatically digitize multiple photos with the same settings, without having to click Scan in your software for each frame. Make sure the software you use also supports batch scanning. With film, batch scanning allows you to insert a strip and have the whole thing scanned automatically and ejected when it's done. With prints, it lets you place prints on the scanner bed one after another without responding to software prompts in between. Some software can also divide images into separate files if you place more than one print on the scanner at once.

Built-in infrared scanning technology to remove dust and scratches

Built-in infrared scanning comes into play if you are scanning color film or a black-and-white chromogenic film. (If your black-and-white film has an orange tint, it is probably chromogenic. See page 71 for more information about scanning black-and-white film.) Kodak's infrared technology is called Digital ICE, and is used by a number of scanner manufacturers. Canon calls its infrared technology FARE (Film Automatic Retouching and Enhancement).

These systems scan film with infrared light to detect dust and scratches, then use software to touch them up. The process is usually very effective and produces excellent results. It is virtually impossible to remove all of the tiny specks that accumulate on film, even if it has been stored and cleaned well, and touching them up by hand is extremely time-consuming and tedious. If you're scanning more than a roll of film, only use a scanner that offers this technology.

If you are scanning Kodachrome film, look for Digital ICE 4, since it offers improvements for handling that type of film. Some scanners offer "digital" dust and scratch removal. This is not based on infrared technology and may not work as well.

Selecting Photos to Scan

Digitizing photos is a time-consuming process, so weeding out some shots before you start can be an effective way to keep your project manageable. If you're scanning slides or prints, you can sort them one by one into a group to scan and a group to put back in the drawer (or throw away, if they're really bad). When you're sorting photos to scan, do it quickly. If you find yourself peering at each image and trying to decide whether it's a keeper, sort your images into three groups: one that you definitely don't care about, one that you definitely want to scan, and one that you might want to scan after you see how scanning the other group goes. If you're scanning photos from albums, there's a good chance that the person who made the albums has already made the selection for you, and they're all worth digitizing. Just leave them in the albums so that you can take them out and put them back during the course of the scanning process.

Film strips require a bit more judgment. The most efficient way to digitize them is to set up your software to scan whole strips at a time with a batch scan process. Unless you're skipping a significant

Tools for Selecting Slides and Film ››

If you're picking slides and frames of film to scan, you can always just hold them up to a bright light to see them. But using a device that will help you get a better view can make the process faster and easier. Here are a few options:

Slide viewer. A slide viewer is usually a little handheld device with a slot in which you insert a slide to see a magnified, illuminated view of it. Such viewers cost about $10 and up and are available from photo specialty stores. Some more expensive models have an automatic feeder for inserting multiple slides, and some have a slot into which you can insert a strip of film to see a magnified, illuminated view of each frame.

Lightbox. A lightbox is a panel or box with an illuminated surface onto which you place slides or film to see them more clearly. Prices start at about $20. Some lightboxes are designed specifically for viewing slides and have angled surfaces with tiers, so that you can look at multiple slides at once.

Homemade lightbox. If you don't want to spend money on a store-bought lightbox, you can create a makeshift one with items that you probably have around the house. Get a cardboard box and cut out or fold in the top and bottom flaps. Then cover one end with a sheet of thick white paper or, if you happen to have one, a piece of translucent white plexiglass or plastic. Tape the paper or plastic to the box to secure it. Set the box up over a light bulb so that the white surface is illuminated from beneath. Make sure it's clean and doesn't get too hot from the light.

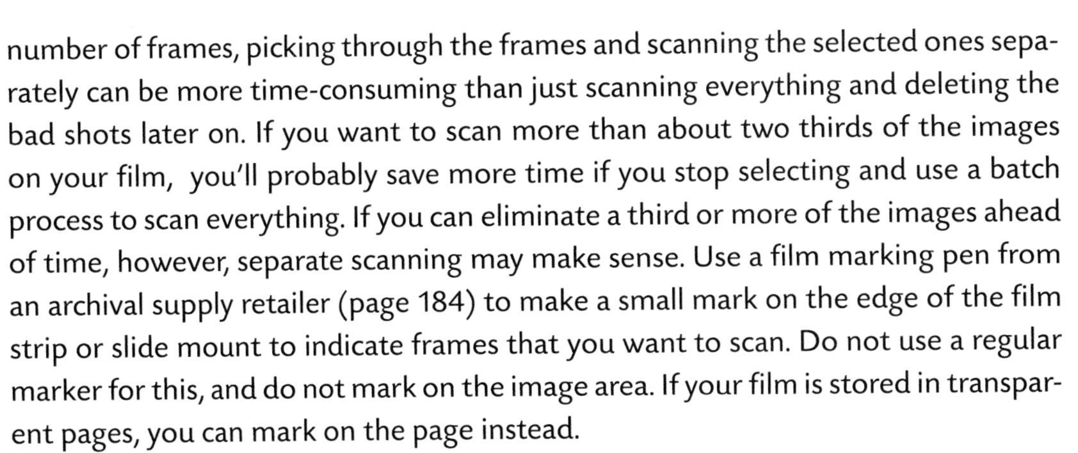

Loupe or magnifying glass. A magnifying lens can show you an enlarged view of your image on a lightbox. Loupes can often stand on their own over the film and sometimes have built-in lights.

number of frames, picking through the frames and scanning the selected ones separately can be more time-consuming than just scanning everything and deleting the bad shots later on. If you want to scan more than about two thirds of the images on your film, you'll probably save more time if you stop selecting and use a batch process to scan everything. If you can eliminate a third or more of the images ahead of time, however, separate scanning may make sense. Use a film marking pen from an archival supply retailer (page 184) to make a small mark on the edge of the film strip or slide mount to indicate frames that you want to scan. Do not use a regular marker for this, and do not mark on the image area. If your film is stored in transparent pages, you can mark on the page instead.

If your stored film is well organized enough that you can quickly sort it into separate rolls and slide batches, put it into film storage pages before selecting the images you want to scan. This allows you to do two things: First, you can easily look at a whole page of images at once on a light box, and second, you can mark the images to scan on the sleeve holding the film instead of the film itself. You can purchase film storage pages from the retailers listed on page 184.

To make the photo selection process more efficient, here are some key criteria that qualify images for the do-not-scan pile:

Irrelevant content. If you simply don't care about what the picture shows, skip it.

Awful composition. Shots of the top of someone's head and unremarkable landscapes with tilted horizons fall into this category. These are the ones that clearly didn't capture what the photographer intended.

Blur. Unless you think there might be some artistic reason for the blur, don't bother with very blurry photos. If you can see that a shot is blurry in a frame of film or a slide, it will be a total mess when you enlarge it.

Multiple shots of the same subject. This is a tricky one if you're looking at film, because there might be important differences between the shots that you can't see until you enlarge them. The photographer might have retaken the photo to get it sharper or make sure everyone's eyes were open. But if you've got a whole strip of pictures of your niece's childhood pet hamster, maybe she'd be happy having just one in the family archive. Use your best judgment here.

The Scanning Process

The process of scanning photos varies depending on the scanner and software, but these steps take you through the process in general terms—and explain the important settings to select.

✔ **1. Clean your work space.** Make sure the area is clean and free of dust. If your photos are in boxes or albums, dust them before bringing them to the work area.

✔ **2. Set up a light.** Adequate illumination is important so that you can make sure your flatbed surface and film or prints are clean. A gooseneck lamp works well. Don't shine a bright light on your monitor, since that may interfere with monitor calibration and your ability to see an accurate image preview.

✔ **3. Clean your scanner.** Use a soft anti-static brush or cloth to remove dust from your flatbed scanner glass, and clean it with a scanner cleaning solution and a microfiber cloth if necessary. You can purchase these items from a photo specialty store or an archival supplies retailer (page 184). A monitor cleaning solution will also work for cleaning the scanner glass, but don't use household cleaners or alcohol-based solutions.

✔ **4. Clean your photos.** Clean your prints, film, or slides by using a soft anti-static brush to remove dust. If the film looks smudged, use a film cleaning solution such as PEC-12 and film cleaning pads to clean it. Put a few drops of the solution on a pad, then wipe it gently along the film

in one direction. You can purchase film cleaning supplies from a photo specialty store. Clean each print or piece of film right before you scan it.

✔ **5. Determine the film type.** If you're scanning film, look at it to determine whether it is positive or negative film. Positive film shows the image naturally, as it would appear in a print. ◀ Slides are always positive, and if the name of the film ends in "chrome," it is probably positive. Black-and-white negative film has the dark and light tones reversed, and color negative film shows the color-wheel opposites of colors, so that reds appear cyan, greens look magenta, and blues look yellow. Identify the film brand by looking for the name along the edge of the film strip. If your black-and-white film has an orange tint, it is probably chromogenic and should be scanned as color film.

✔ **6. Insert the film or place the print in the scanner.** Follow your scanner's instructions for inserting film into a slot or adapter so that the right side is up. Put prints face-down on the flatbed scanner glass, aligning them with the edges.

✔ **7. Open the scanning software.** If you're scanning film, your software may require you to launch the program after the film is inserted in order to detect the images.

✔ **8. Select a film format if necessary.** If you're scanning film in a format other than 35mm, you may need to select the format in your software before scanning. Frames that are smaller than 35mm are usually APS. Frames that are 2.25 inches wide are medium-format film.

✔ **9. Select the photo type and film brand.** For scanning film, select positive (also known as slide) or negative in your software. There may be a separate option for Kodachrome. If the option is available, select the brand and name of the film. For both film and prints, select color or black-and-white, which may also be called monochrome or grayscale.

✔ **10. Select the scanning resolution.** As a rule of thumb, scan prints at a resolution of at least 300dpi and film at a resolution of at least 2000dpi. Higher resolutions produce larger files, but also capture more detail and let you print scanned images at larger sizes. More than 600dpi is generally overkill for prints, but scanning film at 3000 or 4000 dpi can be beneficial, because film contains more detail. Use Wayne Fulton's online calculator (www.scantips.com/calc.html) to figure out which scanning resolution to select to print photos at specific sizes.

✔ **11. Select the scanning bit depth.** The bit depth is the number of bits used to represent each pixel, which in turn determines how many colors (or shades of gray, for black-and-white photos) the scanner can capture. Bits are the zeroes and ones that make up digital data. If your software gives you an option, 8 bits is usually a good choice. In color mode, some software describes this as a 24-bit setting, since color images use three color channels and 3 x 8 = 24. An 8-bit black-and-white scan can capture up to 256 shades of gray, and a color scan with three 8-bit channels can capture up to 16,777,216 colors. If you're going to do a lot of creative editing or just want optimal color fidelity, you can select a higher setting. However, that will significantly increase the file size and scan time, and not all software supports images with high bit depths.

✔ **12. Turn on infrared dust and scratch removal.** If you're scanning color or black-and-white chromogenic film with a scanner that offers this feature, select Digital ICE,

FARE, or whatever name your software uses to refer to infrared touch-up technology. Unless the film is heavily scratched, start with the lowest level of correction. This software is quite sensitive.

✔ **13. Preview the first frame.** Click Preview or Prescan in your software. Some software lets you select the preview resolution separately from the scanning resolution. Using a lower resolution can speed up the process.

✔ **14. Check the frame and orientation.** Rotate the frame and adjust the frame border to fit the image if necessary.

✔ **15. Select automatic exposure adjustment.** Using automatic settings works best for batch scanning, and for single frames until you become more experienced with advanced manual settings.

✔ **16. Select a color space.** If you are given an option, sRGB is usually best. The color space in which that the image is saved is used to describe the palette of colors it is using, so that a monitor can display it accurately or a printer can reproduce the right colors. Computer monitors and commercial printers use sRGB, so saving your images in that color space will allow you to see accurate previews before you order prints of your newly scanned files. Adobe RGB is a good alternative for advanced photographers who know their way around image editing software and will be doing most of their printing themselves on inkjet printers.

✔ **17. Set up and start batch scanning.** Select the batch scanning option for scanning film or a series of prints, and choose a file type to save the images in. The options are generally JPEG and TIFF. Read more about file types on page 60. Most scanning software will allow you to select a scheme for naming the series of image files (for example, Yosemite 0809 1, Yosemite 0809 2, and so on). Your software may also allow you to automatically add metadata such as keywords to your image files as they're scanned and saved. If you're using a film scanner, set the software to eject the film after each strip is scanned. You might also be able to turn off the image display generated after each scan in order to speed up the process. Once you've made all of your selections, click Scan.

✔ **18. Clean the next print or strip of film.** To make the process go more quickly, prepare the next print or piece of film while the scanner is still working on the one before it.

✔ **19. Review the files.** After the first image or batch of images has been scanned, check the resulting image files to make sure everything looks good before proceeding. Does the exposure look balanced, with visible details in both shadow and highlight areas? Does the color look natural? Is the file size acceptable? Make sure that your settings are being applied to all of the images in the batch. If you're scanning film with infrared touchup technology, enlarge your view to 100 percent and make sure that specks and scratches have been successfully eliminated. If not, you may need a higher level of correction.

✔ **20. Put the scanned originals away.** As soon as you take your print or film out of the scanner, put it away so that it doesn't gather dust. If your photos were stored in a disorganized way, use this stage of the process to organize them into new albums, transparent pages, or organizing containers as you scan. Read more about storing film and prints on page 104.

That probably seems like a lot to do, but remember that many of these steps only apply to the initial setup. Once you have your settings applied and get a good batch-scanning rhythm going, the process will speed up.

Once you have made a few scans, you can take a look at the file sizes, multiply them by the number of images you'll be scanning, and estimate the storage space you'll need for your scanned image collection. Enter this information into the Space Estimate column on your Media Collection Inventory sheet so that you can tally it with other media types there to estimate the total storage space you will need.

Ten Scanning Tips

These tools and techniques can help you get the scan quality you want and make the scanning process go faster.

✔ **1. Calibrate your monitor.** See page 102. It's important to get as accurate a preview of colors and brightness as possible, so that you can make adjustments if necessary. Some high-end scanners also provide tools for calibrating the scanner itself, which can be especially helpful for color accuracy.

✔ **2. Make sure your computer is up to snuff.** Scanning requires lots of processing power and memory, so make sure you check your scanner and software system requirements before you start. The newer and faster your computer, the better.

✔ **3. Open your scanning software on its own.** Some scanning software can be opened separately or from inside your image editing software. The latter option usually takes more juice from your computer and can slow the scanning process.

✔ **4. Leave stuck-on prints on the album page.** If your photos are stuck to album pages, don't try to pull them off and risk damaging them. Put the whole album page on the scanner glass and set a book on top of it if the page doesn't lie flat.

✔ **5. Don't use interpolated scan resolutions.** Unless you want to make a gigantic print of a particular image, there's no reason to use software interpolation to scan at stratospheric resolutions, and they can have a negative effect on image quality.

✔ **6. Use automatic color and shadow enhancement tools.** Some scanners and software offer tools for restoring color to faded images and bringing out detail in dark shadow areas. Applying these tools can lengthen scan time, but the improvements are often well worth it. Try them out on a few photos to see whether you like the results before applying them to a whole batch. Kodak's tools, used in numerous scanners, are called Digital ROC and Digital SHO. Other companies use other names.

✔ **7. Avoid sharpening and noise reduction.** Don't apply these tools unless you're experienced and understand the tradeoffs they make with image quality. These adjustments can be made later, will extend the scan time, and may compromise the quality of the scan.

✔ **8. Use de-skew tools.** Some software offers automatic tools for straightening out your images in case they were slightly out of alignment on the flatbed glass. This will make the process go faster, since you won't have to be as painstaking in print placement or make manual corrections after the scan.

✔ **9. Use multisampling to make images look smoother.** Some scanners can take up to sixteen passes over a single frame of film, then average the results to produce smoother, less grainy images. Obviously, this option is time consuming, but it's an excellent tool for improving images that you want to scan at the absolute highest quality, or that are especially marred by graininess.

✔ **10. Print as you scan.** Some software can automatically print your images as they're scanned in a batch process. This can be an efficient way to create a new album of prints and make the most of the time you spend scanning.

Scanning Resources ▸▸

A few manufacturers have stopped making scanners or gone out of business. If you purchase a used scanner, make sure the maker is still in the scanner business and will provide support for your machine.

Film Scanner Manufacturers
Braun www.braun-phototechnik.de
Nikon www.nikon.com
Pacific Image www.scanace.com
Plustek www.plustek.com

Flatbed Scanner Manufacturers
Canon www.canon.com
Epson www.epson.com
Hewlett-Packard www.hp.com
Microtek www.microtek.com
Umax www.umax.com

Scanning Software
Scanning software comes in the box when you purchase a scanner, but if you're not happy with it, you might find one of these third-party programs more powerful or easier to use. You can download free trial versions from the companies' Web sites and see which one makes the scanning process easiest and most efficient for you.

Hamrick Software VueScan
www.hamrick.com
LaserSoft Imaging SilverFast
www.silverfast.com

Scanner Reviews
Optical quality and other performance differences can affect the speed, scan quality, and ease of use you'll get out of a scanner. Before you buy a new model, consult these sources for professional reviews of scanners.

CNET www.cnet.com
Imaging Resource
www.imaging-resource.com
PC Magazine www.pcmag.com
Popular Photography and Imaging
www.popphoto.com

Understanding Digital Image Formats

There are many, many digital image file formats in the world, but for most people just one or two are relevant. It's important to know a few things about file formats in order to make sure you're preserving your digital images in good condition and will be able to use them years from now, when a new image format regime has taken over. You can tell which format a file is saved in by its extension, the letters that come after the dot, such as ".jpg" or ".tif."

JPEGs

Image.jpg

Most people's digital photos are saved in the JPEG format. JPEG files end with the extension ".jpg." (JPEG stands for Joint Photographic Experts Group, the organization that created the standard.) JPEG is a compressed format, which means that it takes the image data captured by a camera or scanner and squeezes it into a package that is a manageable size. All digital cameras and camera phones save photos as JPEGs, and if you've had your film photos put on a disc, those files are probably JPEGs, too. If you're part of the JPEG-only crowd, there are just a few important things that you need to do:

✔ **1. Use the highest image-quality setting** available from your digital camera. When your camera saves a photo as a JPEG, it discards some image information permanently, and there is no way to get it back. The lower the quality setting, the more information discarded. And the more information you lose, the more the quality of your photos will suffer. Converting a low-quality JPEG to a higher-quality JPEG or TIFF later on will not improve the image, because no conversion can restore the elements of the picture that you lost when it was saved originally. The only time when you should choose a lower-quality setting is when you're running out of space on your memory card, since lower-quality JPEGs are smaller. If that happens frequently, it's time to get a higher-capacity card.

✔ **2. Leave your original JPEGs untouched in your digital archive.** To edit a JPEG, make a copy and only edit the copy, leaving the original alone. Because JPEG compression discards some information when the file is first saved, your software must restore that information when you re-open the file. It does that by deducing what the missing pieces should be from the remaining information. For example, let's say your picture shows a patch of blue composed of eight blue dots. The JPEG encoding algorithm might discard two of the dots when it saves the image. When the JPEG decoding algorithm opens the image file again, it looks at the remaining six blue dots, assumes that the now-empty areas should probably be blue as well, and restores the two discarded blue dots. Obviously, a system based on this kind of extrapolation will not restore the missing information perfectly. Each time a slightly imperfect new version of the file is saved, the basis for reconstructing the image the next time moves a little farther away from the original image. That's why it's important to retain the original, unaltered JPEG, so that you can always go back to the first, most accurate version of your picture.

✔ **3. Also use a high quality level when you save JPEGs** in image editing or scanning software. When you save an image as a JPEG with your software, it will sometimes ask you which quality level you would like to use. Don't go below the top few levels. In Adobe Photoshop, a level of 10 or higher is best. There often isn't much of a quality difference between the top two or three levels available relative to the file size increase they cause. If you're concerned about hard drive or other storage space, saving JPEGs at a level or two below the top is usually fine. Only use a very low quality level when you're saving a copy for use on the Web or in e-mail, and its quality isn't of great importance.

TIFFs

Image.tif

If you have scanned photographs, they may be saved as TIFF (Tagged Image File Format) files. Some advanced digital cameras can also save photos as TIFFs. This type of file is very stable and makes a good archiving format, but it is also very large. TIFFs are either uncompressed or use a type of compression that does not cause the image to deteriorate if the file is opened, edited, and resaved. If you have images saved as TIFFs, just leave them be.

RAW Formats

_image.NEF _image.XMP

You have the option of saving your photos as RAW files if you have a digital SLR (single lens reflex) camera or an advanced point-and-shoot digital camera that offers the feature. This is an advantage because RAW formats give you much more control over the look of your photos and preserves more image data than JPEGs. A RAW file is more like a digital negative than a JPEG, because you can adjust its exposure, color, and other image parameters to a much greater degree before displaying the final version of the image. Although RAW files require you to process images with desktop software before you print or share them, they expand your creative horizons and your ability to fix mistakes.

RAW files, however, present a host of problems for archiving. Most camera models have their own proprietary RAW format for recording images. For example, Canon cameras in general use the Canon RAW format, but the Canon Rebel XT uses a different version of the format than the Canon 40D. Canon makes image process-

Image Conversion Programs ▸▸
fCoder Group Image Converter Plus
www.imageconverterplus.com
Mystik Media AutoImager
www.autoimager.com
Online-Utility Image Converter
www.online-utility.org/image_converter.jsp
ReaSoft ReaConverter
www.reasoft.com

ing software that can open and edit the RAW formats used by all of its cameras, and since Canon is a major camera company, most image editing software made by other companies supports Canon's RAW formats if it supports RAW files at all.

But things change. Over the past several years, some companies that were once considered major SLR manufacturers have closed their doors, been bought by other companies, or simply stopped making digital SLRs. Finding software that can open the RAW files produced by their cameras will only become more difficult as time goes on. Fewer and fewer software companies continue to support the obsolete formats as they come out with new versions of their programs, and older programs that can open the files may not run on newer computer operating systems. Eventually, people could find they have collections of photo files that they simply can't open. Since the market is unpredictable, anyone who uses RAW files is risking ending up in that situation.

So what to do? The best RAW archiving option available now is Adobe's DNG (Digital Negative) format. Adobe created DNG specifically to address problems with RAW file archiving, and support for the format has grown in the photo industry. Several camera manufacturers have started using it as their cameras' native RAW format, and numerous software programs can open and process DNG files. Your camera's manufacturer may go out of business one day and stop producing software that can open its RAW files, but the DNG format is still likely to enjoy wide support.

To archive your RAW files, use software that can save copies of the files in the DNG format, and create a DNG archive of your photographs. Look for a batch conversion feature so that you can archive large groups of photos as DNGs at once. If your image processing software doesn't offer these features, try Adobe's own DNG Converter software. You can download it from the Adobe Web site at www.adobe.com.

Handling Mystery Files

If you have files that you can't identify or open, and you think they might be photos, don't just delete them. (If your image files are accompanied by a second set of files that have the same names as the images but a difent extension, read about RAW image metadata on page 87.) Here are the steps you can take to open the files or convert them to a format that you can use.

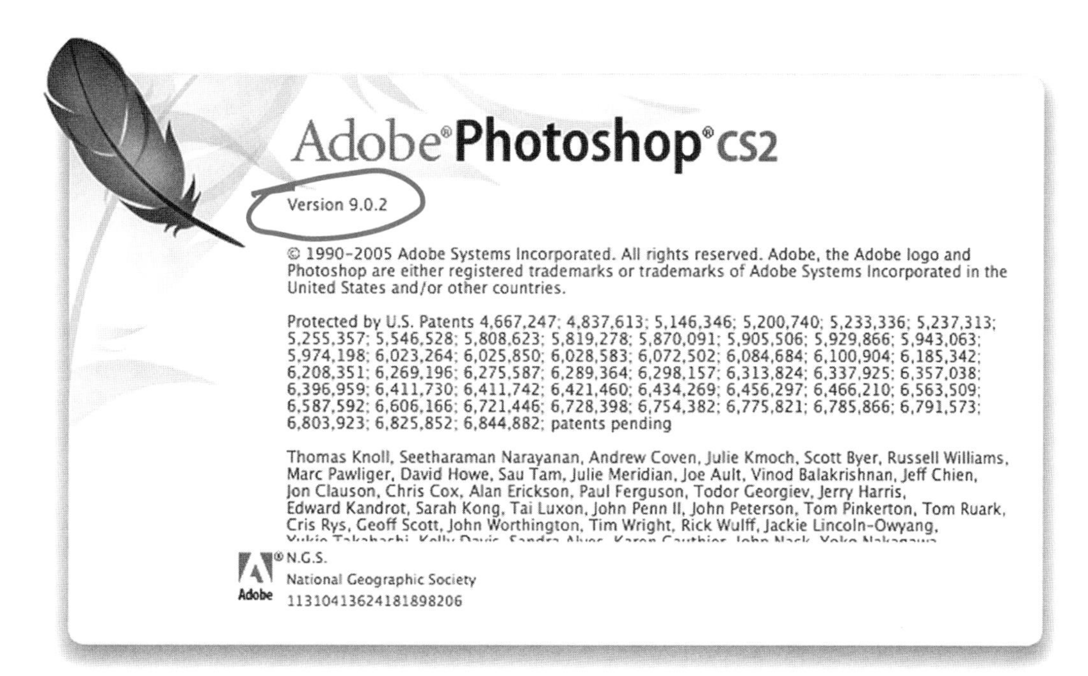

✔ **1. Update your image editing software to be sure you are using the latest version.** Software makers add support for new file formats to their programs periodically. Most image editing software can be updated on a computer connected to the Internet, either through the software's Help menu or through the Web site of the software company.

✔ **2. Try to open the files with the software that came with your current or past digital cameras.** If you already have image editing software installed on your computer, you may not have seen a need to use the software that came with your camera and just left the CD in the box. If you can dig up that CD, install camera maker's software, and update it to the current version, it may provide support for your mystery files that your other image editor doesn't. If you can't find the disc, check the camera manufacturer's Web site to find out which software came with the camera and whether you can download it.

✔ **3. If you are using Windows Vista or Windows XP with the latest updates,** and are trying to view files that you think are in a RAW format, go to the Web site of the manufacturer of the camera that captured the images, then download and install a Vista codec (the term is short for coder/decoder). You should be able to find the codec in the support area of the manufacturer's site. Once installed, it will give Windows Explorer the ability to show thumbnails and previews of RAW image files from the camera, as well as information about the images.

✔ **4. Try a dedicated image conversion program.** Look for one that can convert a whole batch of files at once.

Restoring Damaged, Dirty, and Faded Photos

If you're digitizing old film and prints, it's likely that some of them don't look so hot anymore. The main culprits that make older photos look bad are color fading, color casts, dust, and physical damage. There are many big, thick books that can teach you advanced techniques for transforming your crummy looking old pictures into dazzling photographic wonders that are suitable for a museum. But the goal here is to give you a few simple tools for fixing up a large collection of photographs quickly, so that they're fit for your online gallery, family archive, or album of new prints.

Familiarizing yourself with these tools won't get you a job at the Smithsonian, but you may be pleasantly surprised at how much better your photos look after you apply even the one-click options. You can download free trials of most of the software tools mentioned here.

Restoring Color

To get the best results in color restoration, first calibrate your monitor. (Read more about that on page 102.) Some scanning and image editing software comes with a one-click tool for restoring faded colors and removing color casts. If yours doesn't, or you're not happy with the results, here are some other options:

Kodak Digital ROC (asf.com). ROC stands for Recovery of Color, a task this software performs in one click. It is available as a feature in some scanning software, and can also be downloaded and added to compatible image editing programs as a plug-in.

onOne PhotoTune (www.onOnesoftware.com). This software plug-in lets you tune up the color in photos by viewing improved versions of a photo side by side and selecting the one you like best. It includes a SkinTune tool for adjusting the color in photos by making skin tones look natural, which is useful for restoring color to portraits. This is also a very intuitive tool for adjusting color in photos from your digital camera.

If you are restoring color to a small number of photos and don't want to purchase specialized software to do it, here are three simple steps that you can take in just about any image editing program to restore faded images and remove color casts:

✔ **1. Do an initial automatic adjustment.** Apply the software's automatic tool for adjusting exposure, color, and contrast. This type of tool goes by different names in different programs. For example, it's called Auto Levels in Photoshop Elements and One Step Photo Fix in Paint Shop Pro Photo. In Google Picasa, the tool is I'm Feeling Lucky.

✔ **2. Set a white point.** Find a tool that lets you click on an image to set a white point. In Photoshop Elements, for example, look for the white point eyedropper tool in the Curves palette. Use the tool to click on something in the photo that should be pure white. The border of a scanned print is usually a good place to click.

✔ **3. Boost the saturation.** Locate a saturation setting in your software's image adjustment menu. There is usually a simple slider that you can use to lower or increase the saturation, or vividness, of the colors in your photo. Bump it up just until the colors look vibrant and natural. Don't go overboard.

Removing Dust Spots and Scratches

If scanning your old photos on film is an option, I strongly recommend you do so, and use a film scanner with an infrared dust-and-scratch removal system such as Digital ICE.

If you only have prints of your photos, or are scanning black-and-white film, you'll need to remove dust spots and scratches with software after the images are scanned. Some scanning and image editing software includes an automatic dust and scratch removal function. Try it out on a test image to make sure it is removing specks without removing image details. Also make sure it isn't smudging the specks too much to get rid of them—that can sometimes look even worse. If the automatic tool works well, it will probably serve as good first step before a manual touch-up, but it's very unlikely to do the whole job for you.

To manually remove dust spots and scratches from a photo, open the image file in your software, enlarge the view to at least 100 percent so that you can see each flaw clearly, and go over the image section by section. Here are some tools you can use:

Cloning tools. Many image editing programs have a tool to paste some element of an image over a flaw to cover it up. To use a cloning tool, you select the tool, click to choose a part of the image near the flaw, and then click on the problem spot to paste in a copy of the bit you just selected. The software will smudge the pasted-in element to blend it in. Some cloning tools, such as the Healing Brush in Photoshop Elements, automatically select an adjacent area to clone when you click on the flawed area, so you only have to click once. Cloning tools are excellent for touching up dust spots and small scratches. To avoid conspicuous corrections, make sure the size of the tool is only a tiny bit larger than the dust spot. Some software has linear cloning tools for touching up scratches. They work just like spot cloning tools, but they give you a tool for selecting a linear area over the scratch to paste the clone into.

ImageTrends DustKleen (www.imagetrendsinc.com). DustKleen provides an especially efficient way to touch up dust spots. It performs an initial automatic pass, then lets you finish the job by clicking on spots with an automatic cloning tool. It also lets you quickly remove "corrections" that were mistakenly made to image details instead of dust.

Alien Skin Image Doctor (www.alienskin.com). Alien Skin Image Doctor is a software plug-in that is especially useful for fixing heavy scratches and tears, but it also has dust removal tools. To use it, you draw an outline around the damaged area with your image editing software's selection tool, then click on the Image Doctor to repair it.

Cleaning Dirty Pictures

No, not those. This section is about prints and negatives that have found their way into the hands of small children or other messy types, and have gunk and debris stuck to their surface.

Cleaning Prints

If you have prints with debris stuck to the surface, not just the usual dust, the best way to handle them is probably to scan the image, use image editing software to fix the resulting image, and reprint it. Trying to scrape off the gunk will probably only cause more damage.

If you have a print that is really a mess, but for some reason is too important to just throw away, you can try washing it. Immerse it in a tray of distilled water at about 68 degrees Fahrenheit, and leave it there for a while so that whatever is stuck to it can soften up. Try to brush the debris off gently. When the print is as clean as possible, let the water drip off, then clip the print to a clothesline by one corner or lay it flat in a dust-free area to dry.

Do not try to wash inkjet prints or prints that seem brittle, delicate, torn, or old.

Cleaning Film

If your film has debris adhering to it or there are pieces of film stuck together, buy a wetting agent from a photo specialty store and mix up a diluted solution with distilled water at about 68 degrees Fahrenheit. Follow the instructions on the packaging to get the right dilution. You should only need a tiny bit of the wetting agent. Kodak Photo-Flo is a popular option, and companies such as Ilford and Edwal also make wetting agents. You can buy a bottle for less than $10.

Put your dirty film in a container of distilled water at 68 degrees Fahrenheit and let it sit for a while, until the gunk has softened. Then gently brush the debris off the film. Once you have removed the dirt, rinse the film in the wetting agent solution, and clip a clothespin to each end of the film strip. Hang it up to dry in a place where it won't be exposed to dust. The wetting agent will help it dry quickly and without streaks.

Don't try this with film that seems especially old or brittle—take it to a professional photo lab for restoration instead.

A Word on Sharpening and Noise Reduction

Many old photos have lost some detail and may also look grainy. To make them look a little sharper after scanning them, you can use a sharpening tool that is available in your image editing software. Use a light hand. Oversharpening can make the photo look worse. Most image editing software also offers some type of noise reduction to make photos look less grainy, and there are numerous dedicated programs for noise reduction. Again, overapplying this technology can make photos look worse. Noise reduction often works by blurring the elements of the photo that make it look grainy or speckled with colors. Too much of it can cause your photo to lose detail and look blurry. When you apply sharpening or noise reduction, enlarge your photo to 100 percent magnification on the screen and look it over afterwards to make sure that your changes haven't caused any strange or undesirable effects.

Zero Minute Photo Touch-ups

Damaged, dusty, and faded photos require special attention, but many people's other scans and digital camera photos can often benefit from a little touch-up. Most image editing software offers an automatic image enhancement function that can improve exposure, color, and contrast in one click. But if you just don't have the time to fiddle with photographs before you print them or upload them to an online gallery, Tribeca Labs (www.tribecalabs.com) has an inexpensive piece of software for you. It's called Photobot, and it will automatically enhance your digital photos without you having to do anything at all.

To use it, you install the software, then set it to scan your whole computer or just a designated folder. That's it. The software will find all of your JPEGs and automatically correct poor exposures, remove red-eye, and adjust color and contrast. Every time you download new photos to your computer, it will automatically detect and enhance them, without your having to touch the software. Photobot also preserves the original files, so if you don't like its results, you can restore the unenhanced photos.

Photobot generally does an excellent job with exposures and eliminates most cases of red-eye. Its color results tend to be on the warm side, especially in shots with incandescent lighting. You can put a small sampling of your photos in a folder, download a free trial of the software, and run it on that folder before purchasing it to see if you're happy with the results. Make sure you like the way that skin tones of people in the photos come out before letting Photobot take charge of your whole collection.

Using a Scanning Service

The Goal:

Get all of your prints, slides, and film scanned and saved in a digital format, so the photos can be integrated into a digital image collection, shared online and on disc, and archived for safekeeping.

What You Need:

Money and a few shipping boxes.

Time Required:

Eons less than it would take to scan all of your photos yourself. (All you have to do is pick a scanning service, place an order, and pack your photos.)

Using a scanning service to transfer your prints, slides, and negatives to a digital format is a no-brainer if you have a large collection of them—and most of us do. Reputable scanning services use high-quality equipment, including film scanners that cost thousands of dollars, and employ technicians with skills that it takes significant time and effort to acquire. Above all, a scanning service will save you the very substantial amount of time and the tedious, repetitive work that it takes to digitize a typical family or personal photo collection.

For most scanning services, you ship your photos to the company, where they are scanned and saved as digital files. Then the company ships the photos back to you, along with the digital files on disc. Some companies also let you drop off and pick up your photos in person if you are in the area.

Scanning services fall into three broad categories:

Services that use auto-fed machines to scan many prints at high speed Autofeed services can fulfill scan orders in days instead of weeks and are very inexpensive, with prices as low as about 5 cents per 4x6 print. They usually offer manual services at a higher price for photos that need custom work. One drawback to this type of service is that damaged prints or prints with any adhesive cannot be run through the scanner; if they are put through by accident, they may leave residues that mar the later scans. Also, image corrections are done automatically, which may not be optimal for all photos.

Services that digitize prints by hand on a flatbed scanner and operate solely in the United States or Canada U.S.- or Canada-based manual services are generally the most expensive option, with prices starting at about 40 cents per 4x6 scan. Having each scan processed by a technician instead of a machine can result in higher-quality scans, depending on the skill of the technician and the image corrections required. Manual scanning services take weeks to complete a job, although some offer rush services.

Services that digitize prints by hand on a flatbed scanner and have scanning labs outside of the United States or Canada Services with labs outside the U.S. and Canada operate just like other manual-scanning services, but can charge lower prices because the labor cost for the skilled technicians they employ is lower. Their prices for each scan of a 4x6 print start at about 30 cents, and at least one of them, **ScanCafe,** allows customers to preview the scans online and pay only for those they want to keep (there is a 50 percent minimum purchase). These services may also offer film scanning and custom image-correction services at lower prices.

Some companies that operate labs in the United States use their location as a selling point, but there is no evidence to suggest that there is an advantage in quality or shipping security to using a lab located in the U.S. Whether you are sending your photos across town or to the other side of the globe, make sure that the service you use has secure shipping and tracking procedures and can answer any questions you might have about them.

All reputable scanning services use dedicated film scanners to digitize negatives and slides. Most have Nikon scanners equipped with Digital ICE dust-and-scratch-removal technology. Digital ICE works by scanning the film with an infrared light to capture an image that shows only the dust and scratches; the undamaged film and the image on it are transparent to infrared. The ICE software then uses the infrared image as a map and automatically retouches the scanned image.

Most services send your image files to you on CDs or DVDs when they ship your photo collection back to you. Some can send your files on a USB flash drive or hard drive at an additional cost. A few scanning services offer online archiving or will store your digital files for a specified number of years for a fee. That can be a good safeguard against losing your images through a disaster or theft.

Ten Ways to Get the Best Results from a Scanning Service

✔ **1. If you have prints and negatives of the same images, have the negatives scanned** instead of the prints. There is more detail in the negatives, and the negatives are probably preserved better, too, since most people keep them stored in a dark place. If your prints look faded and have weird color casts, you may be surprised at what great-looking photos you'll be able to get out of—and print from—the negatives.

✔ **2. Choose the right resolution.** Keep in mind that as digital display technologies improve, it will be an advantage to have more detailed images. Have your prints scanned at a minimum resolution of 300 dpi (dots per inch) and your film scanned at a minimum of 2000 dpi. If you opt for 600 dpi scans for the prints, you will capture a bit more detail and be able to make larger prints from the digital files (probably up to 8x10 inches). Scanning prints at resolutions higher than 600 dpi is usually overkill. Scanning film at a high resolution is a good idea, because there are more nuances to capture in film than in prints. If you can afford it, get your negatives and slides scanned at 3000 or 4000 dpi. Note that higher resolutions always mean larger image files.

✔ **3. Compare the basic service** (the one without additional fees) offered by different companies and check what is included. Look for color and contrast correction, red-eye removal, restoration of color fading, image rotation, and cropping of slide mount and print borders. Dust and scratch removal should be standard in color film and slide scanning.

✔ **4. Find out what level of compression** the scanning service uses when it saves scans as JPEG files. What Photoshop calls level 10 compression is optimal. At lower levels, compression artifacts may start to show in the image files. Higher levels generally increase file size without creating a very appreciable improvement in image quality. At the level 10 setting, a typical 4x6 print scanned at 300 dpi will be saved as a file that is roughly 1 megabyte. Most services will also save images as TIFF files for a higher fee. (Read more about image formats on page 60.)

✔ **5. Always pack your photos in a box when you ship them** and use a shipping service that has a package-tracking system. Envelopes and soft packages can get caught in the conveyers used by shipping companies, which can result in their being pulled off the line and being more likely to get lost. To be on the safe side, wrap your photos or albums in plastic or put them in a giant Ziploc bag inside the package. That will prevent them from being exposed to moisture if the package gets exposed to bad weather in transit.

✔ **6. Decide whether you're more flexible with time or money.** To save time, you can send whole albums or projector carousels full of slides to some services. They often charge an additional fee for handling these items, but you may find it worthwhile to pay for the convenience of simply putting your album in a box and shipping it instead of removing all of the photos first. Alternatively, some companies provide shipping boxes that you can fill up with prints that the service will scan for a flat fee with an auto-fed machine.

✔ **7. Find out how your scanned image files will be organized** when they are sent to you. Some services group the files in the same way the materials were grouped when you sent them. For example, boxes, envelopes, or albums will be organized into separate groups. Some companies offer a keywording service, which adds the keywords you request to your image files (more on keywords on page 84). And some will organize photos into separate folders or use custom file names for an extra fee.

✔ **8. Find out the quality of disc your image files** will be saved on (read about disc quality in chapter 1). Some companies offer a higher-quality disc at a slightly higher price. You can always copy your images to higher-quality discs if you're not satisfied with what the service provides, but receiving them on archival quality discs in the first place will save you time and money. DVDs are generally a more economical choice than CDs; although the per-disc price will be a little higher than for CDs, your images will fit on far fewer discs.

✔ **9. If you opt for a service that uses an auto-fed scanner,** go through your prints before you send them and pull out those that have anything on the back or front surface that might stick to or leave a residue on the scanner. Also remove any prints that are torn or otherwise damaged in a way that might cause them to get caught in an autofeeder. Find out if the service can scan those prints manually if you separate them from the rest of the prints (they will likely charge you a higher rate and might take longer). If not, you will need to find another way to scan these images.

✔ **10. If you're worried about shipping losses, break up your collection** into several packages and send them separately. The chances of one package being lost by a reputable scanning service and shipping company are extremely low. The chances of more than one package from the same customer being lost are virtually nil. Having a small portion of your images scanned first also lets you make sure you're satisfied with the scan quality before sending the rest of your collection.

Black-and-White and Kodachrome Film Scanning

The Digital ICE technology that is used to automatically touch up dust spots and scratches during film scanning does not work on black-and-white film because the film blocks infrared light. Kodachrome also poses problems for Digital ICE. For this reason, scanning services charge a higher rate to scan black-and-white and (sometimes) Kodachrome film and retouch the results manually.

Happily, there are exceptions to the rule. Some black-and-white films, called chromogenic films, are composed of the same kind of materials as color film, and can be processed with Digital ICE. If you are sending chromogenic film in for scanning, ask whether the service provider can process it with Digital ICE at the same rate as for color film. Look along the edges of your film strips to find the name of the film.

Common chromogenic black-and-white films include:

-Ilford XP1
-Ilford XP2 Super
-Kodak Professional BW400CN
-Kodak Professional Portra 400BW
-Kodak T400CN
-Konica Monochrome VX-400

Premium-Quality Scanning

You may have a few special photographs that you would like to pass down as family heirlooms or historical artifacts. If you'd like to preserve them in a digital format at the highest quality possible—and be able to make large prints from the digital files—have them digitized on a drum scanner by a professional lab. Negative film and slide film work best on these scanners. A decent flatbed can still scan a print.

Drum scanners cost tens of thousands of dollars, require extensive training to operate, and are generally considered to provide the highest quality of image scanning. You should expect to pay a minimum of $20 per image for drum scanning, and prices can run to hundreds of dollars for large-format negatives. Choose a lab that caters to professional photographers and has an experienced technician working the drum scanner. An excellent resource for finding a pro lab in your area is the Ace Index at www.acecam.com. Follow the Photo Labs link to the geographically organized Custom Photo Labs and Digital Imaging Service Bureaus section.

Scanning Services ▸▸

BritePix www.britepix.com
Digital Memories digitalmemoriesonline.net
Digital Pickle www.digitalpickle.com
DigMyPics www.digmypics.com
FotoBridge www.fotobridge.com
Gemega Imaging www.gemega.com
Larsen Digital Services www.slidescanning.com
My Special Photos www.myspecialphotos.com/
Pixmonix www.pixmonix.com
ScanCafe www.scancafe.com
ScanMyPhotos www.scanmyphotos.com

Using a Scanning Kiosk ▸▸

Photo kiosks are often found in drug stores and large retail chains; you may have used one to print images from your digital camera or camera phone. Some new kiosks made by Kodak include Rapid Print Scanners. These scanners can digitize a pile of photos and put the scanned image files on a CD or create a DVD slideshow in just a few minutes. In addition to correcting color and contrast automatically, the kiosks provide some basic image-editing options such as red-eye correction. If there's a kiosk with a Rapid Print Scanner in your area, it can provide a quick, affordable way to digitize your print collection without requiring you to ship your photos anywhere. Just keep in mind that it won't offer the custom services and premium-quality image file options that most scanning services do, and it won't scan film or slides.

There are some kiosks with integrated flatbed scanners for digitizing prints one at a time. These are not useful for scanning large numbers of photos, and they don't provide all of the creative controls that using your own flatbed and a good software program would. However, a kiosk flatbed can come in handy if you just have a few prints that you want to integrate into your digital collection or make additional prints quickly.

▸▸ Scanning Cost Estimate Worksheet

This worksheet helps you compare the prices and services of different scanning companies. Fill out a separate copy of the worksheet for each type of media that has a different base fee (for example, prints, negatives, slides, black-and-white negatives). Using this worksheet won't give you the precise amount your scans will cost, since it does not include shipping and disc costs for which the companies may charge extra, but you can use it to estimate costs and find a service that fits into your budget.

COMPANY NAME	NUMBER TO SCAN*	BASE FEE**	ADDITIONAL FEE FOR TIFF	CUSTOM FILE NAME FEE	HANDLING	TOTAL COST
Example Company	500	29¢ per image	9¢ per image	9¢ per image	$10 per album	
Subtotal		(500 x 29¢)=$145	(500 x 9¢)=$45	(500 x 9¢)=$45	(2 x $10)=$20	$255
Subtotal						
Subtotal						
Subtotal						
Subtotal						

*Enter the number of printed images, slides, or negatives that you tallied in your **Media Collection Inventory** (page 11).

**Note that some companies charge discounted prices per scan when you send them a larger number of prints. Make sure you take this into account when you enter the base fee.

Creating a System for Organizing Photos

The Goal:

Create an organizational structure for your digital photo files, put your existing images in it, plan how to file new photos, and choose tools for managing them.

What You Need:

Your computer, and any hard drives, devices, discs, or memory cards in which you have stored image files.

Time Required:

About ten minutes to set up the system; additional time to transfer your photos into the system and select photo management tools.

A digital photo collection lets you do things that were simply impossible in the days of film photography. You can bring together the digitized versions of your older film photos, the image files from your digital camera, and your camera phone shots in one integrated archive—and use a wide range of tools to enhance, distribute, and display them. But to take advantage of all the neat things you can do with your digital collection, your collection has to be organized so that you can find photos easily, make sure they're archived properly, and efficiently sort and group them. If this describes the way you and your friends already handle your digital photo collections, please, contact us and tell us more about what life on your planet is like. Here on Earth, most people's digital photo collections are a real mess.

The fact is the organizational state of your digital photos probably has little to do with how adept you are with technology or what kinds of software, cameras, and computers you use. Digital photo organization hinges on the same things that make managing anything succeed or fail: time, interest, realistic expectations, and the organizational structure you use (or don't use). If your photo collection is in disarray, it's probably because you usually don't have the time to organize it—and when you do, you avoid the task because it's just no fun. The more you put it off, of course, the more time-consuming and less fun organizing your photos becomes. The solution? Set up a system for keeping your photos organized that requires as little time and attention as possible, so that it will be easy for you to use it consistently.

It shouldn't take you more than about ten minutes to set up a basic organizational system for your photos. Once you've done that, promise yourself to use it for all of the photos you take from now on. Next, take short chunks of time when you can and integrate your existing photos into your system in small batches.

Be realistic about whether you will perform the tasks required to maintain your organizational system over time, and look for ways to automate any tasks that you think you'll put off. You'll find plenty of suggestions for ways to automate downloads, uploads, photo enhancement, and archiving tasks throughout this chapter and in Chapter 1. Keep in mind that the time and effort you've put into managing your photo collection in the past is a much better indicator of how things will go in the future than the enthusiasm you feel for photo management right now with this book in your hands.

To get down to brass tacks, your organizational system will have two important elements: a folder structure and a set of management tools that are provided by a software program or file browser. Folders are nice because they can give you an organized, visual overview of your images when you view them in a file browser such as Windows Explorer or Mac Finder. When it comes time to archive photos, to upload them to an online destination, or to move them to another computer or hard drive, you'll know exactly where the relevant groups of photos are and be able to select the right folders quickly. But looking through folders is a very inefficient way to find individual shots, especially if you have a lot of images. In order to locate, sort, and group your pictures more easily, you need management tools that let you describe, search, and catalog them.

Before You Get Started

Before you start putting your photographic house in order, make sure that your computer's operating system is up to date. The most recent operating systems give you more tools for viewing, sorting, and managing your image files.

If you're running Windows XP, install the latest service packs and updates if you haven't done so already. Also add the following photo-related tools to your system. They can be downloaded from the Download Center on Microsoft's Web site:

- Photo Info
- Microsoft RAW Image Thumbnailer and Viewer for Windows XP (if you shoot RAW photos)
- Windows Live Photo Gallery

If you shoot RAW images and use Windows XP or Vista, go to your camera manufacturer's Web site and find a free RAW image codec for Windows in the site's support section. Download it and install it. This will allow you to view thumbnails and other information about RAW files in the Windows Explorer file browser and Windows Live Photo Gallery.

Mac users who want to organize their photo files into folders should consider purchasing a file browser to use instead of Finder, which is not designed to offer the kinds of folder and file management tools that Windows Explorer provides. Another option is to use a photo file browser such as Adobe Bridge, which comes with Adobe Photoshop software. Here are a few Mac file browser options:

- RAGE Software Macintosh Explorer www.ragesw.com
- Rixstep Xfile rixstep.com
- Kai Heitkamp Xfolders www.kai-heitkamp.com

Setting Up Your Folder Structure

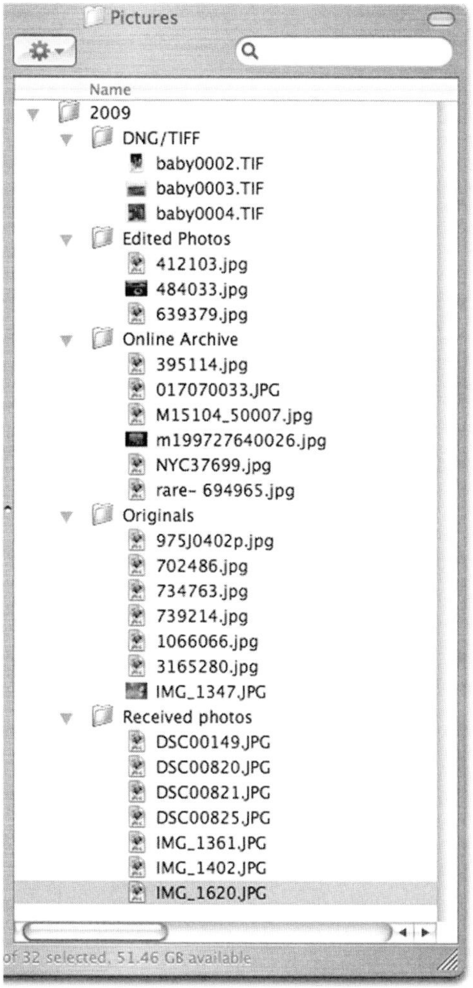

You can create your own variation to meet your specific needs. Here is an example of a simple folder structure for storing all of your image files on your computer. To set up your folders on a Windows system, use the Explorer file browser (this is your desktop browser, not Internet Explorer) or Live Photo Gallery. On a Mac, use one of the file browsers listed above. To start, create new folders going back one year and forward one year.

Originals. Download images from your digital camera and camera phone to this folder and set up any automatic downloading systems to download photos here. Also save scanned images here.

Edited Photos. When you use image editing software to alter your photos, save the edited image files here, and leave original versions in Originals as archived versions. If you use new software or techniques to edit your photos in the future, you'll most likely want to start with the full original image data, not the edited image.

DNG or TIFF Archive. If you shoot RAW photos, you can create DNG copies of them and

Organizing Options ▸▸

- Using your **Pictures** folder as an umbrella folder is a good idea because many photo software programs automatically look for your images there.
- Using a **chronological file structure** with an umbrella folder for each year makes sense. It's an intuitive way to organize things, and all of your digital photo files have dates attached to them, so you can search and sort them by date to put them in folders.
- If you take **more than a thousand photos per month**, you might find it useful to create a subfolder for each month instead of keeping all of your photos for the year together.
- If you take **many photos at specific events**, you might create separate folders named for the events, and include them in the same tier as your monthly folders.
- Read more about Metadata and Tagging options on pages 84-85.

save them in this folder. Here's a space-saving tip: When you convert to DNG, select the option to embed your original RAW files in the DNGs, then delete your RAW files at the end of the year or whenever you archive older photos. You can extract the RAW originals from the DNGs with Adobe's free DNG converter when you want to work with them. If you have decided to archive RAW or other images as TIFFs, you can keep them here, too.

Online Archive or Sync. If you don't want to archive all of your photos online, you can copy a selection to this folder to upload (or set up an automatic uploader to pull photos from this folder). If money, online space, and bandwidth are not issues of concern, you may not need this folder. Instead, you can simply have your uploader pull from your Originals or Edited Photos folder.

Received Photos. If you receive photos from friends and family via e-mail, you may want to save them here, separate from the photos you take yourself.

Organizing in Ten Simple Steps

Once you have set up your folder structure, follow these steps to get your whole collection of digital image files organized.

✔ **1. Designate the appropriate folders in any automatic photo download or editing systems you use.** For example, if you use an Eye-fi card (see page 95), set it to download to Year/Originals.

✔ **2. Designate the folders from which your backup, sync, or online gallery** uploading software should pull photos.

✔ **3. Use the folder system you've set up for all new photos from now on.** Do not drop photos in random places on your computer or make exceptions to your system. If you put your photos somewhere else, it should be because you have made a thoughtful change to your organizational structure.

✔ **4. Starting with the most recent photos, search your system for image files** by year or month, depending on how you've set up your folder system. Search on the "date taken" with your file browser.

✔ **5. Sort out your own photos from other images** that come up in your search, and that you don't want in your collection. These might be images downloaded from the Internet, for example, or images used by software programs you've installed. To separate your images from the rest, sort them by criteria that differentiate them, such as file type or camera model. To do this in Windows Explorer in XP or Vista, go to View / Choose Details or View / Sort By and select your sorting criteria.

✔ **6. Set your file browser to show thumbnails or previews of your images** and delete any photos you don't want to keep. These may include very blurry shots, repetitive shots, and photos of things that you just don't care about. Look at photos at full size to judge their quality, not just the thumbnails.

✔ **7. Move (don't copy) your existing photos into the folders you've created,** starting with the most recent ones. You probably will find edited photos mixed in with originals. That's fine. Just put them all in the Originals folder for now.

✔ **8. If you're an advanced photographer and shoot RAW photos,** add a step for creating a DNG archive folder of your original RAW files at this point. Use a batch converter to handle all of the RAW files in your Originals folder quickly.

✔ **9. Rinse and repeat.** In other words, keep searching and moving files in chronological increments, going back in time, until you scrape the bottom of your collection. If you don't have time to do this all at once, take 15 minutes when you can to add another year's worth of photos, or a few months, to your organizational structure. When you've organized all of the photos from the past year, create a new folder structure for the previous year. Keep working back in time until you have filed all of your photos.

If you're a person who accumulates very large digital photo collections, you may run out of room on the drive where your My Pictures or Pictures folder is stored, especially if you use a high-resolution camera or save RAW files and TIFFs. For photographers in that situation, it makes sense to archive older photos on a separate hard drive or storage device. If you do so, be sure to use exactly the same folder structure for the archive folders as for the ones in your My Pictures folder. That way, when you transfer older folders from the My Pictures folder to the archive in the future, they will fit right in. If you've been squirreling away photos for years on multiple hard drives, devices, and discs, consider bringing them together in a

Timesaver ▸▸
An alternative approach is to bypass the process of creating a folder structure and simply select a powerful photo management program that can quickly search, sort, and manage all of the image files that are scattered around your computer and any external hard drives. This can save you the time involved in setting up a folder structure and sorting your existing photos into it, but it will also mean that if you stop using the software for any reason, your photo organization system will disappear. I don't recommend this approach for large photo collections.

Renaming Your Photos ▸▸

Another useful step in organizing your photos is giving your image files meaningful names, instead of the names they have when they come out of a camera or camera phone.

You should incorporate renaming into your download process going forward. For existing image files, you can use the batch tools in a Windows file browser to rename photos as you move them into your new folder system. However, many image editing and management programs provide more flexible tools for renaming files in batches, so you may want to wait until you've selected your management tools. If you do choose to rename your images with your file browser, here is the process you can follow in the Windows Explorer file browser:

- Select the group of photos to rename.
- Type a new name into the highlighted file name. Make sure you retain the file extension (e.g., ".jpg").
- Press enter on your keyboard. Windows will rename all of the selected files with the new name you typed in and sequential numbers.

unified archive on a single high-capacity storage device (read more about storage options in Chapter 1). This will allow you to organize, search, and manage your photos much more easily.

✔ **10. When or if you have time, sort out the edited photos** that are in your Originals folders and move them into the Edited Photos folders. This is an optional step; you may decide that it's not worth your time to go back and sort your photos in this way, and simply keep your originals and edited images sorted in the future. One way of identifying photos that have been edited is to look for File Date and Modified Date information that is different for the same file. You should be able to see this in your file browser. Some advanced photo management software will also note the name of software used to alter an image in the image file's EXIF metadata.

Choosing Photo Management Tools

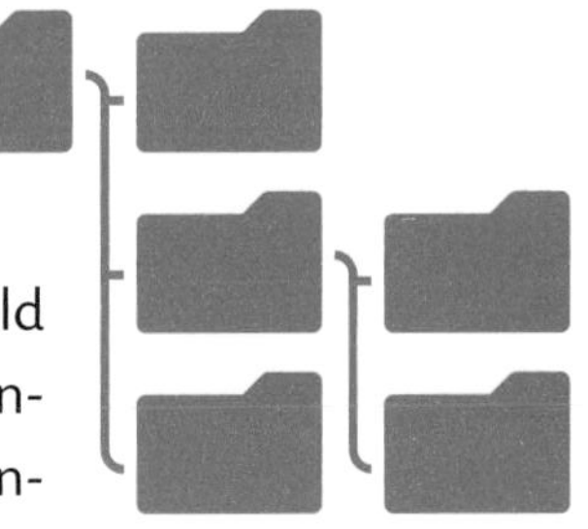

Once you've organized your photos into a simple folder structure, you can choose a program that will give you the tools you need to manage your images. Your choice should take into account the price you're willing to pay for image management software and the level of detail you want to use in managing your photos. Your options include:

- Free tools provided by your operating system's file browser and photo software that came with your computer or digital camera
- Photo management tools integrated into your image editing or photo workflow software
- Dedicated photo management software

If you just want to keep your photos reasonably organized, you may find all the tools you need in the file browser and the image software that came with your computer. These include tools for searching, sorting, batch renaming, and even viewing and adding EXIF and IPTC metadata. Snapshot photographers who won't use more than a few simple tools and don't spend much, if any, time editing photos, only need the tools provided by iPhoto and the Finder file browser on a Mac or by Windows Live Photo Gallery and the Windows Explorer file browser on a PC.

Whichever program you choose to manage your photos, you should stick with it. Don't use more than one software program or other resource to organize your images unless you understand how the programs you're using work in tandem. Using multiple programs in a random way to organize your photos will probably only increase the disorder.

Ten Key Photo Management Tools

When you're selecting the software to manage your photos, you need to know which tools are most important. Basic programs and file browsers may offer just a few of them, whereas advanced programs will provide dozens of useful photo-management tools. Before you buy a program, download a free trial version from the software maker's Web site, and make sure you find it straightforward and intuitive to use. A program that combines advanced features with simple wizards and tutorials, and lets you select a basic or advanced level, will accommodate you best over time. Here are some of the most useful photo management features to look for:

✔ **1. Flexible batch renaming.** All of your software choices will probably offer some file renaming tools. Look for those that let you select large batches to rename at once and give you numerous options for sequencing the included files. For example, if your batch will be renamed "Yosemite xxxx," you should be able to select sequential numbers or date and time information in a variety of formats to fill in the "xxxx" information for each file. An example of what you might choose for vacation photos from July 2009 would be "Yosemite 0709 1," "Yosemite 0709 2," and so on.

✔ **2. Strong metadata features.** Choose a program that lets you add tags to images and save the tags as standard EXIF or IPTC metadata. Advanced photographers should use a program that also allows viewing and editing full metadata lists. See page 84 for more information on metadata.

✔ **3. Efficient searching and sorting.** Make sure the program you choose offers intuitive, efficient ways to search for photos, group them, and sort them by date, name, caption, photographer (sometimes called author), keyword, and other characteristics.

✔ **4. Ranking and filtering tools.** These are tools that let you mark photos as favorites, or rate, label, or flag them. Once you've done so, you can filter your photos to find only your favorites or all of those with a particular rating, label, or flag. Ranking and filtering tools can be used to sort photos within the program and keep them organized while you're editing them or choosing photos to distribute.

✔ **5. Powerful download and export options.** Look for downloading tools that let you automate tasks such as file renaming, tagging, and DNG creation so that they occur whenever you transfer photos to your computer. (Read more about downloading options on page 92.) Built-in tools for exporting photos directly to online galleries, including them in e-mail, and saving them to discs or hard-drive archives can also make it quicker and easier to keep your photos organized.

✔ **6. Cataloging.** Some programs allow you to create catalogs of images. (Catalogs are sometimes called "libraries," "projects," "albums," or some similar term.) A software catalog works the same way as a paper or card catalog, in that it holds information about a group of items, and it lets you organize that information in a way that is helpful for you. But when you move things around in the catalog or add information to it, that doesn't affect the actual items that are listed. They're still just sitting in the warehouse, or, in the case of image files, sitting in the folder where you downloaded them. The advantage to using an image catalog is that it lets you pull up and sort groups of photos quickly; easily make changes to batches of images; and keep groups of photos organized as you edit them. Catalogs can be especially useful for advanced photographers working with large numbers of images, and many advanced image editing programs use catalogs to organize photos within the software. However, catalogs can only be used within the program where you created them. And, as noted, moving images between catalogs is not the same thing as moving the actual files in a file browser.

✔ **7. Integrated file browsing.** In addition to general file browsers such as Windows Explorer, some photo management and editing programs let you browse, move, and copy image files in a file browser interface. This differs from using a cataloging system, because you can move and change the files themselves, instead of just transferring information about them between catalogs.

✔ **8. Deduping tools.** Some advanced programs offer tools that automatically find duplicate files or files with the same names, and let you delete or rename them. This is a much more efficient means of eliminating duplicates than picking through files manually.

✔ **9. Backup tools.** If you choose to back up only your image files instead of everything on your computer, or if you edit photos frequently and want to make sure your images are safe between regular system backups, a program with built-in backup tools may be your solution. Look for a program that lets you set the backups so they occur automatically.

✔ **10. Offline file cataloging.** Some advanced programs have the ability to include image files that are archived on a disc or external hard drive in their photo catalogs. This will allow you to see your entire collection all at once in a thumbnail browser.

Ten Photo Organization Rules to Follow

Here are some guidelines to follow in order to keep your photos organized once your system is in place.

✔ **1. Make sure your digital camera's date and time are set correctly.** This will ensure that your image files can be sorted chronologically by date. Your Internet-connected computer and camera phone are likely to have their date and time set correctly automatically, but check scanned and mobile images to ensure they are dated accurately as well.

✔ **2. Never dump photos onto your desktop or into a random folder.**

✔ **3. Don't overcomplicate things.** Software that offers rich photo management features can be great for serious photographers who shoot and need to manage thousands of photos. But if you're just a snapshot photographer, don't try to use complicated tools that will require a lot of time to learn how to use.

✔ **4. Rename photo files when you download or scan them.** That's the best time to change file names from something like _923427.jpg to an identifiable name such as Yosemite Vacation 1.jpg. Use a batch renaming process so that it only takes a few seconds to handle a whole memory card full of files.

✔ **5. Add metadata when you download or scan photos.** That's the best time to attach tags, keywords, and other metadata to your shots.

✔ **6. Delete bad photos right away.**

✔ **7. Make DNG archive copies when you download RAW photos.** If you're an advanced photographer shooting RAW images, creating DNG copies to save as archived originals is an excellent way to ensure that your photos are preserved. Use a program that can automatically generate DNGs as your RAW files are downloaded, or can do a batch DNG conversion as soon as you download.

✔ **8. Rely on an automatic backup system.** Your digital photo collection should always be backed up. You can read more about backup options in Chapter 1. Setting up a system that automatically backs up image files is the best way to go for most people. Ask yourself: If you were already in the habit of regularly backing up your photos manually, would you be reading this book? Be realistic, save yourself some time, and automate.

✔ **9. Keep your photo collection unified.** These days, many people have multiple devices that can store digital photos—laptops, portable media players, external hard drives, and the list goes on. That makes it easy for a photo collection to start sprawling over multiple devices, which is a direct path to chaos. Make sure you download your recent photos to your unified folder structure, instead of letting them sit on a laptop or in your camera phone.

✔ **10. Straighten up your photo collection at the end of the year.** Put a date on your calendar each year—or each month if you take a lot of pictures—as a time to organize your photo collection. If your digital photo collection is large enough that you can't store it all on your main computer hard drive, this is a good time to move the photo folder for the previous year to the storage device where you archive older portions of your collection. Also bring any files that may have gone astray back into your organizational system, set up a new folder structure for the coming year, and take care of any neglected tasks, such as renaming files, deleting bad photos and duplicates, and adding metadata. Finally, ask yourself how well your system has worked for you and make adjustments to your workflow. Try to automate tasks you neglected.

Organizational Resources ››

For Snapshot Photographers

Free Editing and Management Tools

Apple iPhoto (included in the iLife suite that usually comes with a Mac, but can also be purchased separately) www.apple.com
Google Picasa picasa.google.com
Kodak EasyShare www.kodak.com
Windows Vista: included Explorer & Photo Gallery www.microsoft.com
Windows XP: included Explorer & downloadable Live Photo Gallery www.microsoft.com

Image Editing Software with Management Tools

Adobe Photoshop Elements with Adobe Bridge www.adobe.com
Arcsoft PhotoImpression www.arcsoft.com
Corel MediaOne www.corel.com
FotoFinish www.fotofinish.com
Nova Development Photo Explosion Deluxe www.novadevelopment.com
Roxio PhotoSuite www.roxio.com

Photo Management Software

ACDSee Photo Manager www.acdsee.com

For Advanced Photographers

Photo Editing and Workflow Software with Management Tools

Adobe Lightroom (RAW workflow software) www.adobe.com
Adobe Photoshop CS3 with Adobe Bridge www.adobe.com
Apple Aperture (RAW image workflow software) www.apple.com
Bibble Labs Bibble Pro (RAW image workflow software) www.bibblelabs.com
Light Crafts LightZone www.lightcrafts.com
Phase One Capture One (RAW image workflow software) www.phaseone.com
Corel Paint Shop Pro X2 www.corel.com

Photo Management Software

ACDSee Pro 2 Photo Manager www.acdsee.com
Arcsoft MediaImpression www.arcsoft.com
Breeze Systems BreezeBrowser Pro www.breezesys.com
Camera Bits Photo Mechanic www.camerabits.com
Cerious Software ThumbsPlus www.cerious.com
Extensis Portfolio www.extensis.com
FastStone Image Viewer www.faststone.org
FotoTime FotoAlbum www.fototime.com
IDimager www.idimager.com
IrfanView www.irfanview.com
Microsoft Expression Media www.microsoft.com
More on page 182

Tags, Captions, Ratings, and Other Metadata

Image files aren't just representations of your pictures in the form of digital data. They're really whole packages of information. In addition to holding data that represents the photo itself, an image file contains lots of information *about* the image. All of this additional information in your image files—whether it's added by you, your camera, or a piece of software—is called the metadata, the image itself being the data. And metadata is the key to keeping your images organized.

Some of the metadata is attached to each image by the camera that takes it. Camera-generated information includes the date and time when the photo was taken—if your camera's clock and calendar are set correctly, an essential feature to check. It also includes details about the settings that were used to take it. Software programs sometimes add to this information when you make changes to images. You yourself can contribute more information about an image with software tools or through an online photo gallery.

You can think of a digital image file as an envelope that contains a picture (the data) with sticky notes attached to it (the metadata). Each sticky note has a standardized list of items on it, and when you, your camera, or your software add metadata to the image file, you are filling in the items on the list.

EXIF and IPTC are the two most commonly used types of metadata in digital photography. EXIF stands for Exchangeable Image File Format, and IPTC is named after the organization that developed it, the International Press Telecommunications Council.

EXIF Metadata

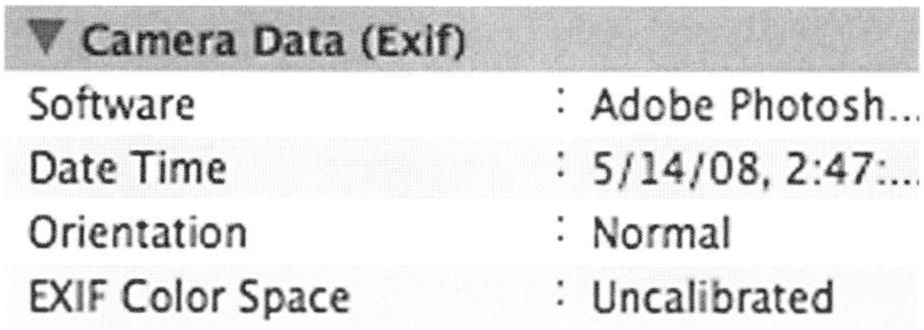

▼ Camera Data (Exif)	
Software	: Adobe Photosh...
Date Time	: 5/14/08, 2:47:...
Orientation	: Normal
EXIF Color Space	: Uncalibrated

When you take a picture with a digital camera, it enters the date and time, and lots of information about the settings the camera used, on the image file's EXIF list. Your software or online gallery service uses the EXIF metadata attached to your images to help you organize them. For example, when you use software to sort your photos by date or organize them in a calendar, the software will look for the date of each image in the EXIF metadata. In some cases, image management software also adds keywords and other basic information to the EXIF metadata when you type in the information in the software.

The only case in which you would want to change the EXIF metadata generated by your camera would be if the date and time on your camera were set incorrectly. If so, you can correct the date and time in the EXIF metadata after you've downloaded your photos to your computer. To do this, you'll need image editing or media management software that provides a tool for changing a photo's date, or allows you to open and edit EXIF files.

IPTC Metadata: Tags, Captions, and More

You can use IPTC metadata to customize the way you organize, sort, and search for your images. It's a long list of items that are left blank so that you can fill them in yourself with software or an online gallery tool. The items can include the name of the photographer, captions, keywords, and copyright information, to name just a few options.

Software for casual photographers and some online gallery sites provide tools to add IPTC metadata to your photos by adding tags, keywords, captions, or descriptions. Tags and keywords are words that can be used to search for images and sort them into groups; captions and descriptions include lengthier information that provides insight into each image but won't function primarily as an organizational tool. Each image can be assigned multiple tags and keywords.

When you add a tag, keyword, caption, or description to your photo this way, the information is entered on the IPTC list so that it stays stuck to the photograph when you move it around. So, for example, if you add the tag *Yosemite* to a photo in your media management software and then upload the photo to your online gallery, it will still be tagged as *Yosemite* on sites that support IPTC. Wherever you take that picture, you'll be able to find it by searching for the *Yosemite* tag.

More advanced image editing and media management programs give you full access to your IPTC lists, letting you add many types of information.

Tips on Adding Tags and Keywords

The most important organizational tools that metadata make available are keywords and tags. Here's some advice on choosing them.

Choose terms that reflect what is most relevant to you about the photograph. For example:

- names of people, places, and events
- the type of photograph, such as *landscape* or *portrait*
- the activity, such as *dancing* or *soccer*
- the type of object or creature, such as *house* or *blue-footed booby*

Don't be so specific that each image has a different tag, and don't be so general that the tags fail to differentiate your photos in a meaningful way. Remember that you will use the tags later to find groups of photos that have something in common.

If you do use very general tags, add more specific ones too, so that you can sort photos into subgroups. For example, if you tag a large group of photos with *family,* you might also tag the images with the names of the individuals shown. That way, you can not only find your entire family collection but also locate photos of specific people without scrolling through the whole family collection.

Don't enter time and date information, since it is already attached to the photograph in the EXIF metadata.

Look for software or an online service that lets you tag images in batches, so that, for example, you can select ten photos and type the word *Yosemite* once to tag them, instead of typing *Yosemite* ten times.

Also look for software that lets you incorporate IPTC tagging into the process of downloading photos to your computer. That way, you can set the software up to automatically add the tag *Yosemite*, for example, to all of the photos you transfer to your computer after your vacation. Then you can go in and add more specific tags to each photo if you have the time and inclination.

The All-Important Test Run

Be sure to try out the software or online tool you'll be using to add descriptive infor-

Metadata and RAW Files »

If you shoot RAW image files with a digital SLR or advanced compact camera, the metadata may be stored in sidecar files. Sidecar files are just what they sound like—little files that are attached to the main image files and travel along beside them as they move from place to place.

There's just one big thing you have to know about sidecar files: *You should not delete them*. Sidecar files usually have the extension .xmp. If you look in a folder of RAW files and see a bunch of files that have the same names as the image files but with different extensions (for example, photo1.dng and photo1.xmp), just leave them be. If you delete them, your metadata will be lost.

mation to your photos and confirm that it works the way you expect. The goal of this test run is to make sure it's entering the items on the IPTC or EXIF list in a standard way that will make the metadata readable in other software and sites. Unfortunately, that's not always the case. If your software isn't using a standard metadata list, the information you've entered could be detached from your images when you move them around.

Add sample tags, keywords, captions, or descriptions to a test photo using the software or online site in question, then either upload it to your online gallery or download it to your desktop and open it there. The information you've added should show up on the other end. If it doesn't, you'll be better off with a different tagging tool.

Ratings, Favorites, and Categories

Many software programs allow users to give their photos ratings that can be used to sort the photographs. For example, a typical system will give you the option of assigning a photo one to five stars. Some software lets you designate photos as favorites or assign them to predetermined categories such as People or Places.

These features can be useful tools for organizing your photos within a particular program, but they have an important limitation. Ratings, favorites, and categories are usually not added to the images' metadata. If you think of the EXIF and IPTC data as lists on two sticky notes, these additional elements are on other sticky notes that will fall off a picture once you take it out of the software program. Ratings, favorites, and categories are well suited for short-term organizational tasks within a specific program, such as marking photos that you want to print or e-mail as you look through them. For long-term organization that will be retained in another software program or an online gallery, stick to IPTC and EXIF tags and keywords.

Geocoding: Organizing Photos by Location

The Goal:

Add information about where your digital photos were taken, so they can be organized by place or digitally pinned to a map.

What You Need:

Geocoding software or a photo-related program that includes geocoding. A GPS device, a memory card with WiFi-based geocoding, or a camera or camera phone with built-in GPS is highly recommended.

Time Required:

Automatic geocoding takes the time that your software or device needs to process images and location data—usually just a few minutes. Manual geocoding generally requires as much time as it takes for you to drag images to the appropriate spots on a map or type in location information.

Sorting photographs by the location where they were taken is one of the most intuitive ways to organize images. Most of us have old prints with place names and dates penciled on the back, as well as albums and envelopes of photos for particular vacations or travel destinations. Organizing digital images by place is useful, too, especially if you travel frequently or want to know exactly where you took particular photographs—of a rare bird or on a mountain hike, for example. In the digital world, the equivalent to penciling a place name on the back of a print is called geocoding. Geocoding matches an image with location information—usually in the form of latitude and longitude. The most common source for the data is a GPS device, although some systems use cell phone towers and even Wi-Fi signals to calculate and record locations. Geocoding embeds the location information in the image file's metadata so that it can travel with the image file wherever you move it (see page 84 for an explanation of metadata).

An increasing number of photo organizing programs include tools for searching and sorting photographs by location, and even for finding them on a map. Windows Vista can sort photos by place, too. Some photo gallery and photo sharing Web sites automatically pin photographs that are marked with location information to an online map, sometimes combining them with information about the place where they were taken. That gives you and other people a way to view the photos in context and to see pictures that different people took at the same location.

+/- Geocoding Pros and Cons

+ Geocoded photographs can be searched and sorted by location with some software programs and operating systems.
+ Geocoded photos can be digitally pinned to the place where they were taken on an online map. Some online maps will combine them with information about the location.
– GPS devices generally do not capture data indoors, which makes GPS-based geocoding useful mainly for outdoor photography. Some geocoding software will tag images with the location logged at the time that is closest to the photo's time stamp.
– Geocoding usually requires images to be processed by software or to be tagged manually.
– Geocoding with GPS data records the location of the GPS device, so if you use a long zoom lens to photograph a far-off subject, the location attached to the photo will not be precise.

How to Geocode your Photographs

With a GPS device and a digital camera:

Using GPS information is one of the easiest and most precise ways to geocode an image. Almost all software that geocodes images automatically matches the images with GPS data collected at the same time that the photographs were taken. If you don't have GPS data (or use a WiFi-based system—see page 95), you will have to geocode your photographs manually (see below). The GPS information can come from a few sources:

-A camera or camera phone with a built-in GPS receiver
-A GPS logger made specifically for geocoding images
-A standard, handheld GPS receiver that can output data to a computer via USB

✔ Step 1: Acquiring the location information

It's essential to make sure that the date and time on your digital camera are correct. Automatic geocoding works by matching the time recorded in your image file with the time in your GPS data.

If your camera includes an integrated GPS receiver, all you have to do to acquire the location data for geocoding is to go out and take pictures. GPS-enabled cameras are still uncommon in the consumer market, but there are some available.

Most photographers end up using a separate GPS device. Some GPS loggers can be mounted on the hot shoe of an advanced camera, and some are made to be carried in a pocket or worn on the body with an armband or other accessory. To use such a standalone geocoding device, simply turn it on when you're taking photographs and turn it off (mostly to conserve battery power) when you're done. While it is on, it will periodically record GPS data as you move around.

Geocoding Resources »

Geocoding Loggers and Cameras with Built-in GPS

ATP Electronics www.atpinc.com
Dawn Technology www.dawntech.hk
General Electronics www.general-imaging.com
GiSTEQ www.gisteq.com
Jobo www.jobo.com
Pharos www.pharosgps.com
Red Hen Systems www.redhensystems.com
Sony www.sony.com

Online Photo Gallery and Sharing Sites That Support Geocoded Photos
These photo sharing sites have maps to which you can digitally pin your photos, so that you can see them in a geographical context. A few offer geocoding software for computers and camera phones.

Yahoo! Flickr www.flickr.com
GEOsnapper http://www.geosnapper.com
Google Picasa Web Albums http://picasa.google.com/
ipernity www.ipernity.com
locr www.locr.com
Panoramio http://www.panoramio.com/
pikeo www.pikeo.com
SmugMug www.smugmug.com
Klika TripTracker http://triptracker.net
Woophy http://woophy.com
Zooomr http://www.zooomr.com

Automatic Geocoding Software
These programs automatically add GPS data to images, mainly JPEG files. If you're an advanced photographer shooting RAW images, make sure you select a program that can geocode them.

Atomix Technologies JetPhoto Studio www.jetphotosoft.com
Breeze Systems Downloader Pro www.breezesys.com
GeoSpatial Experts (several programs) www.geospatialexperts.com
Houdah Software Houdahgeo www.houdah.com
locr GPS Photo www.locr.com
MMI Software PhotoGPSEditor www.mmisoftware.co.uk
Ovolab Geophoto www.ovolabs.com
OziPhotoTool www.oziphototool.com
More on page 183

Advanced photographers using a GPS-enabled digital SLR camera may prefer a GPS logger with a data cable; the cable connects to a port on the camera and transfers location data as you take each image. There are also accessories that plug into an SLR and receive data from a Bluetooth-enabled handheld GPS device. In both cases, the camera automatically geocodes the images and no additional processing is necessary. The cameras that support this system are generally professional and semipro digital SLRs.

If you want to use a handheld GPS receiver that isn't designed to geocode photographs, make sure that it can record location data while you're taking pictures and download the data to your computer in a file. The advantage of using a device that is not made specifically for geocoding photos is that it will perform additional functions. The disadvantage is that it probably won't come with geocoding software, so you'll need to select a program separately.

✔ **Step 2: Matching the location information with your photos**The way the geocoding process usually works is that you download your images and your GPS data onto your computer, then import them into the geocoding software. If you're using a compact camera with built-in GPS, simply import the images into the software that came with the camera, since the GPS information will already be attached to them. There are also some GPS loggers that allow you to insert your camera's memory card when you're done taking pictures and then automatically geocode your images right in the device. If you're using a handheld GPS receiver to record your location track, the geo-coding software you select will probably convert the data into the right format for geocoding. If your software can't convert the information, you can use a separate tool to convert it into a GPX file that your geocoding software can use. Converters are available at www.gpsvisualizer.com.

Without a GPS device:

If you don't have a GPS device or if you want to add location information to photographs you've already taken, you can still geocode your images. Just select a software program or Web site that allows you to geocode your photos manually. Usually, you will need to drag and drop an image (or group of images) onto a map, or enter the location as text; the software automatically adds the location data to the image file.

With a camera phone:

You can follow the same geocoding process with a camera phone as you would with a digital camera and a GPS device, or you can download camera phone shots to a computer and geocode them manually. However, camera phones offer some additional possibilities.If your camera phone has a built-in GPS function, you may be able to install software on the phone that will automatically geocode your images when you take them and upload them directly to an online map on a photo-sharing Web site. (This assumes you're outdoors with good GPS reception.) Also, some software programs for cell phones with wireless Bluetooth can receive data from a Bluetooth-equipped GPS unit and automatically geocode photos.

Even if you don't have a GPS phone or Bluetooth receiver, there is software available for some smart phones that can extrapolate your location from nearby cell tower locations and automatically geocode and upload your shots. Using cell tower data is less precise than using GPS, however, unless you're standing right next to the tower.

Downloading Digital Camera Photos

Unless you're a professional photographer, you probably don't give a lot of thought to the way that you download photos to your computer. But dropping images into a random folder on your computer whenever your memory card fills up is sort of like tossing your new possessions down the basement stairs every time you come home from the store. Before long, the prospect of trying to find anything down there will only fill you with dread.

Take advantage of downloading tools to keep your images organized from the beginning and you'll save a lot of time and grief down the line. A good media management program—or image editing software with strong organizing features—gives you plenty of tools for organizing your images as they are downloaded. (Read more about those types of software on pages 29 and 83.) At the very least, a program will let you select the folder where you want images to be downloaded every time you connect your camera or insert a memory card.

If you're otherwise satisfied with your media management or photo editing software, but it doesn't offer many options for automating tasks during downloading, consider using Breeze Systems Downloader Pro ▾ (www.breezesys.com). It's a dedicated image downloading program that lets you customize what it does with your photos as it transfers them to your computer.

Useful Tasks to Choose for Downloading

Here are some of the organizational tasks that a good media management, downloading, or image editing program can perform while it downloads your photos to your computer:

- **Rename your image files.** You select a naming scheme, and the software renames your images automatically as they're downloaded. For example, if you select a date and sequence scheme for the file names, your images will be called January 1 2009-1, January 1 2009-2, and so on.
- **Create a new folder for each group of downloaded photos** and assign the folder a custom name. It's a good idea to keep all of your photos in one umbrella folder so that you don't lose track of them, but you may want to separate photos from different occasions into subfolders.
- **Rotate your JPEGs.** Some programs will rotate images if necessary so that they appear right side up.
- **Let you add keywords, descriptions, and other metadata** to the images being downloaded. Adding descriptive information to your photos from the beginning is an excellent way to keep your photos organized. (Read more about metadata on page 84)
- **Geocode your photos.** Some downloading software can attach location information to your images as it downloads them, as long as you load a corresponding GPS track into the software first. (Read more about geocoding on page 88.)
- **Save backup copies of your photos** on a separate hard drive.
- **Set the DPI for the photos.** This does not change the images, but simply attaches information to them that will instruct a printer to print them at a particular dots per inch setting. This can make the process of printing your photos quicker and more efficient.
- **Delete your photos from the memory card** when the download is complete. This saves you the step of deleting your photos from the memory card when you put it back in the camera. Make sure that your downloading system is working smoothly before selecting this option!

Starting Your Download Software Automatically

You can set up most programs to open automatically and start the download process whenever you insert your camera's memory card into a card reader in your computer or connect to your camera or a USB drive. This is a convenient feature if the main reason you use memory cards and USB drives is to transfer photos. But if you frequently transfer other types of files to your computer with a USB drive or memory card, you might find it more annoying than convenient to have the photo software pop up every time you connect a USB drive; use your best judgment here.

Here is how you set the AutoPlay feature in Windows XP or Vista to automatically open the program you use for downloading whenever you connect a camera or put a memory card in your card reader:

Windows XP

✔ **1. Insert a memory card** into your card reader or connect your camera to your computer with a USB cable.

✔ **2. Click on My Computer** in your computer's start menu. Your memory card or camera should appear in the list.

✔ **3. Right click on the icon** for your memory card or camera, and select Properties from the menu.

✔ **4. Select the AutoPlay tab.**

✔ **5. Select Pictures** from the drop-down menu.

✔ **6. In the Actions section,** check the box for Select an Action to Perform, then select the software that you want to use to download image files from the list.

✔ **7. Click OK to finish.**

Windows Vista

✔ **1. Select Control Panel** from the start menu.

✔ **2. Click on** the AutoPlay icon.

✔ **3. Choose the software** that you want to start up when you insert a memory card or attach a camera by selecting options in the drop-down menus.

Apple OS X

If you're using a Mac, you don't have the option—or the necessity—of using AutoPlay. Instead, use a given program's Preferences to set up the program to open when you connect your camera to your Mac via USB or put a memory card in your card reader. Programs for the Mac integrate this function directly into the software interface in a standard way, so that you don't have to use an additional tool. Your Mac will start only the program you last set to open automatically, so that multiple programs don't open when you connect a camera or insert a memory card.

0 Zero Minute Downloads

Eye-Fi SD Cards

If you often forget to download photos from your digital camera until your memory card fills up and you don't have room for more pictures, or if it's just another hassle that you'd rather avoid, there is a solution for you: the ◀ Eye-Fi SD card. The card takes the process of transferring images to your computer—and even uploading them to your online gallery—off your hands completely. With the Eye-Fi system, downloading your photos works like this: Plunk your camera down somewhere in the vicinity of your computer (make sure it's on). Walk away. That's it.

You can take advantage of the Eye-Fi system if your computer has WiFi wireless connectivity and your camera uses an SD memory card type.

An Eye-Fi SD memory card has a built-in wireless transmitter so it can automatically send photos from the card to a WiFi-connected computer. You don't even have to take the card out of the camera for this to work. You simply bring your camera within range of the computer that you have set up to receive images from your card, and the transfer begins automatically.

To set up the Eye-Fi system, you install Eye-Fi's software on your computer, then use it to select the folder where you would like images transferred from your camera to be stored. You can also elect to have the software automatically upload images to your online gallery if the gallery site you use supports this feature (check Eye-Fi's Web site at www.eye.fi for a list of compatible sites).

Eye-Fi's Explore card can also be used at WiFi hotspots when you're away from home. Whenever you're in range of a compatible hotspot, it can upload your photos wirelessly to online destinations directly from your camera. What's more, the Explore card automatically geocodes your photos by using a WiFi-based location information service.

Eye-Fi SD cards are sold by major electronics retailers under the Eye-Fi and Lexar brand names.

WiFi cameras

Another way to download your photos with no effort after the initial setup is to use a WiFi enabled camera. Nikon and Panasonic ▶ makes them, and other companies such as Kodak and Canon have in the past. Check with a well stocked camera store to see what's currently available.

A good WiFi camera lets you set it to automatically download photos to your WiFi connected computer when you're in range of its wireless signal. It may also allow you to set the downloading software to automatically upload your images to your online gallery. The camera should also let you connect to public WiFi networks so that you can wirelessly upload photos from your camera to your online gallery when you're away from home.

Getting Photos Off a Camera Phone

If you're like a lot of people, you take pictures from time to time with your camera phone, but you don't often share them with friends, and you may not have bothered to find out how to get them into your computer or online gallery. Camera phones aren't as standardized as digital cameras, and it can be a hassle to figure out how the photo transfer process works every time you get a new phone. The good news is that there are plenty of simple options for transferring photos from your phone. And as mobile memory cards become more common in phones, the way that transfers are made is becoming more standard from device to device.

Once you know how to transfer photos from your camera phone to your computer or online gallery, you can integrate them into your digital image collection, where they will be easier to organize, print, and share.

Transferring Photos to a Computer

Here's a look at the ways that photos can be sent from a camera phone to a computer. Most camera phones don't offer all of these transfer methods, but all camera phones offer at least one.

Mobile memory card

Many newer camera phones have memory card slots that are just like the ones on digital cameras, only smaller. Common mobile memory card types include MiniSD, MicroSD, and Memory Stick Micro.

You can remove the mobile memory card from its slot and put it in your computer's card reader to transfer images, just as you would a regular size card—with the caveat that you'll probably have to put the camera phone's memory card into an adapter before inserting it in the card reader. Mobile memory cards usually come with the necessary adapters. You can also buy USB card readers that have mobile memory card slots built in, which can be useful if you transfer images frequently from your camera phone to your computer.

If you use a mobile memory card, make sure that you have your phone set up to save photographs to the card, and not to the phone's internal memory.

Note: Don't confuse your phone's subscriber identity module (SIM) card with a mobile memory card. If you use a GSM phone, it has a small, rectangular smart card that probably fits into a slot in the battery compartment. This is the SIM card, which holds information about your telephone service account—not a mobile memory card, where you store photographs. Mobile memory cards usually fit into a slot on the side of the phone instead.

USB cable connection

Some camera phones can connect to a computer via USB. They may come with the required USB cable or may work with a cable that can be purchased separately. Check your phone manual or the manufacturer's Web site to see if you can use a USB cable with the phone. If so, the USB connection should work the same way it does with a digital camera: You connect the phone to your computer with the cable, and the phone shows up as a drive on your computer so that you can transfer the images.

Some phones come with desktop software, which opens and helps you download photos when you connect via USB. If your phone didn't come with software, you can purchase a mobile/desktop synchronization program from a third-party software maker. This type of software not only helps you to download photos, but can also synchronize many kinds of data between your phone and your computer, including contacts, calendars, and music. Some options are listed in the Resources section.

Bluetooth wireless connection

Many camera phones can connect to other devices wirelessly via Bluetooth. If both your phone and your computer have Bluetooth built in, you should have no problem with opening the connection and initiating the transfer. If your computer does not have Bluetooth, you can buy a USB Bluetooth adapter for it. Look for a Bluetooth 2.0 adapter with EDR (enhanced data rate) for the fastest transfers. You can select the folder to which your images will be transferred with your computer's Bluetooth management tools.

If you haven't used a Bluetooth connection between your phone and computer before, you will have to set up the connection first. To do this in Windows, go into the Bluetooth section in the Control Panel and follow the instructions for adding a device. The information about USB synchronization software applies to Bluetooth downloading as well.

IrDA wireless connection

Try the IrDA wireless option only if you don't have a better alternative (and, of course, only if your phone has an IrDA port). It's not as fast or convenient as other methods. If your computer also has a built-in IrDA connection—unlikely, but possible if you use a laptop—then you can simply beam your photos from your camera phone to your computer by aligning the IrDA ports, selecting the images on your phone, and then selecting the option to send them via IrDA. If your computer doesn't have IrDA already, you can purchase an IrDA USB adapter. The software that comes with the adapter should allow you to select the folder to which your images will be transferred.

E-mail or MMS

If you have a camera phone with a data service plan, you can e-mail photos to yourself (and other people, of course) using your mobile e-mail software, or send them to your e-mail address via MMS. E-mail and MMS are usually not the most convenient means of transferring images, and using MMS is generally not a good idea because it reduces the quality of images in order to transmit them. However, these options may come in handy if you're away from your computer and want to make room on your memory card without losing the images you've captured. Make sure you know how much your service provider is going to charge you per message before you make a habit of transferring photos this way.

Sending Photos to an Online Gallery, Social Network, or Blog

One way of managing camera phone photos is to send them directly from your phone to your online gallery, bypassing your computer altogether. If the gallery site you use provides good organizational tools, and you don't care about integrating your camera phone shots into your desktop image collection, this might be a good option for you. You might also choose to send your photos directly to a social network or blog, especially if they're topical shots that you want to share with others quickly.

Many sites provide uploading tools. The site provider will usually give you an e-mail address to send images to via MMS or mobile e-mail. You give the site your mobile phone number or a special password; in a password system, you put the password in the subject line of the message when you send an image from your phone, so that the site associates the image with your account. When you send images in, the site automatically adds them to your gallery, blog, or personal Web page.

Mobile/Desktop Synchronization-Software Makers ▸▸

These companies make desktop software that helps you download your photos from a camera phone to your computer when the phone is connected via USB or Bluetooth, as well as helping to synchronize and manage many other types of data.

FutureDial www.futuredial.com
Mobile Action global.mobileaction.com
MobTime www.mobtime.com
Susteen www.susteen.com

There are also services that provide more convenient, and sometimes less costly, tools for getting your images from your phone to the Web. Some allow you to send full-resolution images instead of the smaller files that MMS can handle. These tools can be used to send images to many different gallery, blog, and social networking Web sites. BlogPlanet (www.blogplanet.net) offers mobile software that lets you blog and upload photos to your blog directly from your phone. Umundo (www.umundo.com) provides an e-mail address to which you can send images from your phone; Umundo then posts the images to the online galleries, social networking site, or blogs that you designate for your account.

Some media management suites, such as SugarSync (www.sugarsync.com) and Transmedia Glide (www.transmediacorp.com) have mobile components that will synchronize the photos on your camera phone with a Web gallery and a folder on your desktop. If you have a compatible phone, all you have to do is set up the software once; it then runs automatically whenever you take new pictures.

Most mobile uploading software and services are free, although your phone service provider will charge you for data transfers (check your plan). Some camera phone manufacturers pre-install uploading software on models with higher-quality cameras. Take a picture with your camera phone and then look at the options that are available after the shot to see if there's an uploader already installed.

Zero Minute Up and Downloads

Here's how you use Kodak Picture Upload Technology to upload photos from your Bluetooth-equipped phone to your computer: Turn your computer on. Set your phone down in the general vicinity of the computer. Walk away. That's it.

To acquire this magical system, you have to purchase a USB Bluetooth adapter (sold by Belkin, www.belkin.com) that comes with the Kodak software. When you install the software on your computer and attach the adapter, the software guides you through the process of setting up your phone and computer to automatically connect and download new pictures. You can also upload to an online Kodak gallery at the same time.

Or you can have your photos automatically uploaded directly to an online gallery of your choice by installing software from ShoZu on your phone. To find it, go to www.shozu.com. All you need is a compatible phone (check the list on the site) and a data service plan. The software is free, and you can install it easily by following the instructions on ShoZu's site.

Once you install the software and set it up to make automatic transfers, ShoZu uploads your shot every time you take a picture. If you happen to lose your connection during an upload, ShoZu picks up where it left off when you get your bars back. If you prefer to be consulted before uploads, you can set up ShoZu that way, too. The software can also automatically geocode images taken with a GPS-equipped phone (see page 88), and it provides tools for renaming images and adding tags and descriptions before they're uploaded.

Printing Digital Photos

Back in the era of film photography—just a few years ago—most of us printed nearly all of the photos we took. In fact, before digital photography, a print of a photograph was practically synonymous with the image itself, and a print was often the only form in which people kept their photographs, the negatives being thrown away or lost over the years.

These days, a print tends to be viewed as just one of many forms a photograph might take. Most of us print only a small fraction of the digital images sitting on our hard drives or displayed in online albums and digital frames. But what many people may not realize is that prints don't just provide a way to *display* images. They can also offer one of the most reliable and long-lasting means of *preserving* photographs.

Digital files may become unusable after new formats are introduced, CDs and DVDs containing photos may become unreadable, and you can count on hard drives crashing. It's not unlikely for those things to happen in less than a decade. But a high-quality print can last hundreds of years. You probably won't want to print all of your digital images, but making hard copies of your favorites is a wise choice.

If you have old, fading snapshot prints with colors that no longer look natural, the wisdom of using prints to preserve photos may not be obvious. The important thing to understand is that not all print types offer the same longevity, and all prints must be stored and displayed properly in order to be preserved (more on that in the next section). Think about it: You might have prints from the 1970s that look pretty bad, while your album of family photos from the 1920s still contains beautiful printed images. The older prints were simply made with materials that remain more stable over time.

With the advent of digital imaging, the issue of print stability became immensely more complex. Now there are many more types and combinations of chemical processes, printers, inks, and papers available to the average consumer, and many people aren't aware of how great the differences in print longevity between them can be. When you're buying a printer or supplies, or choosing a printing service, it's natural to think about cost, convenience, and features instead of wondering how long it will take for the prints you'll make to fade into oblivion.

Overlooking image permanence is a mistake

Wilhelm Imaging Research ▸▸ (www.wilhelm-research.com) is an independent organization that tests printer ink-and-paper combinations to estimate how long they will last under a variety of conditions. Wilhelm artificially accelerates the effects of light and pollutants on the prints it tests, producing data on inkjet, dye-sublimation, and silver-halide photo lab prints. The test results are published online, and Wilhelm's site also provides general information about print preservation.

though, if you want the photos you print to last. Before you buy a printer or use a printing service, consider the printing technology it employs and check with ◀ **Wilhelm Imaging Research** to see if independent print-longevity test results are available. There are three basic digital printing technologies available to consumers these days:

- **Inkjet:** Most home desktop printers and some photo kiosks use inkjet technology. Inkjet printers apply tiny droplets of either dye-based or pigment-based ink to the paper surface.
- **Dye-sublimation:** Portable printers and photo kiosks sometimes use dye-sub technology. Dye-sub printers use heat to apply layers of primary-color dye and a protective laminate to paper. The dye reel and paper pack are sold together in a package.
- **Digital silver halide:** Stores, photo labs, and online services usually use digital minilab machines that combine a digital input process with the same type of silver-halide photo chemistry that is used to print film photographs.

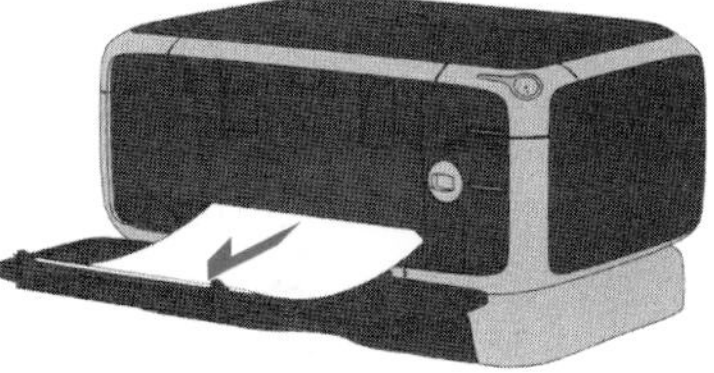

+/- Digital print technology pros and cons

The quality of the materials and equipment used will affect the results you'll get from any print technology, but these are the broad strokes on how your options compare.

✔	INKJET—DYE	INKJET—PIGMENT	DYE-SUBLIMATION	DIGITAL SILVER HALIDE
LONGEVITY	Generally not as long-lasting as inkjet with pigment-based inks	Generally very long-lasting	Generally less long-lasting than inkjet and better lab prints	Varies widely, but generally not as great as the most long-lasting inkjet prints
COST	Varies according to photo content, usually higher than minilab	Varies according to photo content, usually higher than minilab	Fixed , usually higher than minilab	Fixed by retailer, usually lower than with home printing
IMAGE	Generally capable of very vibrant colors	Sometimes not as vibrant as dye-based	Notable for smooth gradations and tonality	Notable for smooth gradations and tonality
PAPER	Ability to print on wide range of paper and other media types	Ability to print on wide range of paper and other media types	Glossy paper only	Many services offer only glossy or matte (but also print on photo gift items)
SIZE	Usually up to 8x10 or 13x19	Usually up to 8x10 or 13x19	Often limited to 4x6 size; larger than 5x7 is rare and expensive	Many services print up to poster sizes
SUPPLIES	Wide variety of papers and inks available	Wide variety of papers and inks available	Supplies available only from printer manufacturer	A reputable lab can tell you what type of minilab and paper it uses.

Ten Home Printing Tips

If you choose to print your photos at home on a desktop or portable printer, the process can get complicated. The most common problem with home printers is that the colors and exposure may not come out in print the way they look on a computer monitor. These tips will help you get the results you want and produce prints that will last.

✔ **1. Calibrate your monitor.** To see the brightness and color of your photo accurately before you print it, you need to ensure that your monitor is set to the appropriate brightness and color settings. To do this, purchase a monitor calibration package. Prices start at about $60. A package includes a small device to put on the front of the monitor and software that guides you through the calibration process, which takes just a few minutes.

✔ **2. Set the appropriate print (or output) resolution.** With a decent printer, resolutions as low as 200 dpi can produce excellent results. Using 240 to 300 dpi is usually a safe bet, and you can go to 600 dpi or higher for nuanced photos or your best prints.

✔ **3. Use high-quality paper.** For the best archival quality, choose an acid-free 100 percent cotton rag or alpha cellulose paper.

✔ **4. Use good ink.** Don't buy cheap inkjet inks sold under office-supply and other third-party brands. They can produce unpredictable results, are likely to fade more quickly, and may even clog your printer nozzles. Stick with the printer manufacturer's inks or a high-quality alternative that has been reviewed and tested for print longevity.

✔ **5. Buy a printer that uses more than four inks.** All photo inkjet printers have at least four inks—cyan, magenta, yellow, and black. The printer mixes the inks in different amounts at each pixel to create a wide range of colors. However, some colors are hard to render with just three color inks. Higher-end printers add more inks to render greater nuance and accuracy in colors.

✔ **6. Rely on automatic tools if you're a novice.** Preparing images to print can get complicated if you're making manual adjustments with image editing software. Fortunately, the automatic tools for image enhancement and print preparation in many software programs are usually quite good. Some printers with built-in memory card slots or printers that can connect directly to a camera also provide automatic image enhancement functions. Look for features for automatically improving exposure, color, and contrast before you print. Then look for automatic layout tools to help you place one or more photos on a page.

✔ **7. Softproof if you're an advanced photographer.** This lets you see an onscreen simulation of how your photos will look when they are reproduced with a specific printer and paper combination. If you're using advanced image editing software, download the ICC profiles for your printer and paper types from the manufacturer's Web site. Then use your image editor's softproofing functions to preview your image before printing. For example, in Photoshop, go to View, Proof Setup, and Custom, and select the profile for your printer and paper from the Device to Simulate list. Select Simulate Paper Color. Selecting Perceptual Rendering Intent usually works well.

✔ **8. Make a ringaround print.** Some image editing and printing software will let you automatically create a print that shows numerous variations of a single image on one page. You can also create one manually by making copies of your image in your software with slight variations in color, brightness, contrast, or saturation, then printing them all at small sizes on one page. This is called a ringaround print because the standard layout is to put the original image in the middle, with the variations in a ring around it. A ringaround allows you to choose the look you prefer before printing it at a larger size, and to minimize the paper and ink you waste on prints you don't like.

✔ **9. Resolve color management conflicts.** If your prints keep coming out with colors that look markedly different from the images on your monitor, make sure that there isn't a conflict between your printer's software and your image editing software. Go into your printer driver software, which you can usually get to from your image editing software by clicking on the Print option and selecting printer setup or properties. Find the color management options, click on ICM if you're using Windows, and disable ICM or select No Color Adjustment. This will turn off the driver software's color management so that you can control color through your image editing software only.

✔ **10. Use the right ink for black-and-white prints.** If you're going to print monochrome photos, look for a printer that allows you to use a "photo black" cartridge, as well as a light gray ink. This will let you make prints that show more detail and tonal range. Advanced photographers can also buy a custom monochrome set with four or more inks from a reputable ink maker such as MediaStreet, MIS Associates, Lyson, Luminos, or Piezography.

Printer Resources ▸▸

Photo Printer Manufacturers

Canon www.canon.com
Epson www.epson.com
Hi-Touch Imaging www.hitouchimaging.com
Hewlett Packard www.hp.com
Kodak www.kodak.com
Sony www.sony.com

Printing Software
Many people find all the printing features they need in their image editing software and the driver software that came with their printer. But if you're looking for something more, check out these programs.

Arcsoft Print Creations and PhotoPrinter www.arcsoft.com
Canon Easy-PhotoPrint www.canon.com (look for it as a download under Canon printer model names in the Support section)
Qimage www.ddisoftware.com/qimage

Monitor Calibrator Makers
ColorVision www.datacolor.com
X-Rite www.xrite.com

Printer Review Sources
Before you buy a new printer, consult these sources for reviews. Look for the print quality and speed, what software is included, and built-in automatic functions that might make printing easier for you.

CNET www.cnet.com
Imaging Resource www.imaging-resource.com
PC Magazine www.pcmag.com
PC World www.pcworld.com
More on page 183

Storing Prints and Broadcasting Photos

As I mentioned in the previous section on printing photos, prints are a great way to preserve your images. To keep your prints in good condition for as long as possible, pay attention to the way you handle them, where you store and display them, and the materials that your photo boxes, albums, and frames are made from. Materials that are suitable for storing photographs are often called "archival," which means that they do not contain or release chemicals that can damage them.

Prints can also be harmed by ultraviolet light, heat, moisture, airborne pollutants, and, of course, rough handling. Exposure to any of these elements can cause fading, color shifts, surface blotches, and physical damage.

To preserve your prints in their full glory, follow the guidelines below. If your prints are currently stored in a bunch of old shoeboxes and magnetic albums, it's time for you to go shopping. (See page 184 for Archival Printing Supply Retailers.)

Do:

✔ **1. Store your prints in closed** albums and boxes that keep out light, and protect framed prints with UV filtering glass. The less exposure to light, the better.

✔ **2. Make sure that storage boxes,** albums, framing mats, and other paper and cardboard storage supplies are both acid-free and lignin-free. To find out for yourself whether a product is acidic, use a pH testing pen. The ink will turn a particular color when you draw a small line on paper that is acidic.

✔ **3. Use album pages,** framing mats, and other paper materials made of 100 percent cotton rag or alpha cellulose.

✔ **4. Use storage supplies** made of archival polypropylene, polyester, polyethylene, or Tyvek.

✔ **5. Store your prints in a place** with a relatively constant temperature of 70 degrees Fahrenheit or cooler.

✔ **6. Store your prints in a place** with no more than 50 percent relative humidity. If you think the humidity might be higher, buy a humidity card for a couple dollars. Patches on a scale printed on the card will turn colors to indicate the relative humidity. Hygrometers provide more precise information, but have prices that start around $40. (Overly low humidity is also a concern; relative humidity below 30 percent can cause prints to become brittle.)

✔ **7. Add silica gel packets** or desiccant cartridges to boxes of photos if you're concerned about high humidity.

✔ **8. Consider keeping** particularly valuable or important prints in a fireproof safe.

✔ **9. Make two prints** of the photos you want to display. Keep one of each in dark storage as an archival copy.

✔ **10. Handle your prints** by touching only their edges. Even if your hands are clean, your fingers will leave a slight oil residue. Or, for more freedom in handling your photos, pick up a pair of white cotton gloves when you're ordering print storage supplies.

✔ **11. Use a pencil,** a specialty photo pencil, or an acid-free photo pen for writing on the backs of prints. Make sure you don't press so hard that you leave an indentation, and don't use a pen that will show through the print.

✔ **12. Purchase materials** that have passed the standard Photographic Activity Test. PAT is an international standard test for determining whether materials meet archival criteria. Products don't *have* to have a PAT approval to be archival, but it's a plus.

Do Not:

✔ **1. Don't display photos** in direct sunlight or in strong fluorescent light. If you use fluorescent lights near your displayed photographs, look for bulbs with a UV filter.

✔ **2. Don't trust every label** claiming a photo storage product is "archival" or, even more ambiguously, "photo-safe." Look for specific information about what it is made of and whether it is acid- and lignin-free.

✔ **4. Don't use adhesives** or laminates on photos, and don't use "magnetized" album pages. If you must use an adhesive, purchase an archival type from an archival product supplier.

✔ **5. Don't use materials made of PVC** or vinyl to store photos. Even if the prints are not in direct contact with these plastics, the gases that the materials emit can cause damage.

✔ **6. Don't store prints in a hot or humid place,** or in areas where the temperature fluctuates. You should not keep photos that you want to preserve in attics, kitchens, basements, shelves or closets on outside walls, or areas near radiators or heat ducts.

✔ **7. Don't store prints on low shelves,** under pipes, near windows, or anywhere else where they might be exposed to moisture or flooding. If you're concerned about possible exposure to water, put your albums and photo boxes into giant polypropylene Ziploc bags.

✔ **8. Don't store or display prints outdoors,** or in a place with the same environmental conditions as an outdoor area, such as a semi-enclosed porch or outbuilding.

Storing Film »

All of the same rules about storage apply to negatives and slides: Keep them cool, dry, and away from light and pollutants, in appropriate containers. Protecting film from abrasive surfaces and dust is also important.

A binder box is a good choice for storing negatives or slides. It's a three-ring binder made of archival materials, but it closes like a box to keep dust and light out. You can fill it with archival transparent pages that hold your negatives and slides. The pages are easy to remove so that you can view their contents near a light, and they have opaque label areas where you can note the contents of the page. Archival slide storage boxes are also a good choice and are very compact.

Don't store your film in a completely airtight container such as a plastic film canister or sealed box or bag; some films release gases as they age, and a build-up can cause deterioration.

Always make sure you dust and clean your film before slipping it into a storage sleeve. For information about how to clean your film without damaging it, go to page 55.

✔ **9. Don't expose your prints to smoke,** chemical or cleaning sprays, or cooking vapors.

✔ **10. Don't store your prints in contact with leather.** Although leather is sometimes used for album covers, it is not acid-free.

✔ **11. Don't add decorations** to photo album pages that are meant to preserve prints. If you want to make a scrapbook or an album that includes decorative elements, use copies of your photos and store the originals elsewhere. It's usually hard to tell whether the decorative items you might use are archival.

Broadcasting Photos on TVs and Digital Frames

Displaying your photos in a digital photo frame or on a television can be a good way to keep a slideshow of your favorite shots in view or show them to a large group of people. You can transfer photos to most digital frames by inserting a memory card or by connecting a camera with a USB cable. Some digital cameras can be hooked up directly to a television, too. Putting a memory card in a gaming console or portable media player that can connect to your TV is another way to get photos onto your set. And systems such as Apple TV, TiVo, and Windows PCs running Media Center can make your whole digital photo collection available on a connected television.

For those far-flung friends and relations in your life who might be computer-averse, here are a few tools and services that you can use to broadcast your photos directly to them without too much effort:

PhotoShowTV

If you (and your viewers) live in an area where the cable-television provider offers PhotoShowTV, you can use it to broadcast your photos in slideshows on a TV channel. Get the free PhotoShow software from the Roxio Web site (www.roxio.com). Then create and name a slideshow of your photos and select the option to make the show available on TV. Cable subscribers in areas with PhotoShow service can choose and watch your slideshow on a video-on-demand channel, just as they would tune into any TV show.

Ceiva Digital Frame

Ceiva (www.ceiva.com) sells digital frames along with service plans that let you send images to a frame from any Internet-connected computer or camera phone. The frame is connected to a phone jack and automatically downloads new images every day from Ceiva's server. If you want to broadcast your photos to someone who doesn't have a computer or Internet connection, this is a good way for them to receive the pictures.

WiFi Digital Frame

A WiFi digital frame, as the name suggests, has built-in wireless network connectivity. To use the WiFi function, you must have an Internet connection and a wireless network set up in your home. Set up the WiFi connection on a digital frame by using onscreen menus to tell the frame to detect and connect to your network. This is usually done with a remote control, but some frames have a touch-sensitive area for navigating menus, which can be a quicker and easier way to control them. What you can access with the WiFi connection depends on the features of the model you buy. Here are some to look for:

Access to your online photo gallery. Find a frame that supports the gallery site you use to host your online photo collection. To see a slideshow of photos from your gallery in the digital frame, go into the frame's setup menu, select your site, and enter the user name and password for your gallery.

RSS feeds. RSS, which stands for Real Simple Syndication, is a system for distributing content on the Web by allowing people to subscribe to content "feeds." Some online galleries automatically generate an RSS feed of the photos in your albums. If your site does this, you will see an icon like the one above in your album. Anyone can click the icon to subscribe to a feed of those photos that you make public. Subscribers will receive your feed through the RSS reader that they use to bring together all of the subscriptions they have, much like their own customized online magazine.

FrameChannel. The online FrameChannel service (www.framechannel.com) lets you create personal "channels" of your own photographs, and also lets you subscribe to channels of photos and other content from professional media outlets—including the National Geographic Society, the publisher of this book. FrameChannel uses the same technology as RSS feeds do, but it has the advantage of letting you create, subscribe to, and organize your feeds, or "channels," in one central account online. To see the channels on a digital frame, you or your viewer need to make sure it is a WiFi frame that supports FrameChannel.

Choosing Photo Galleries and Communities

Online photo gallery and sharing sites are excellent resources for displaying and archiving photographs. If you've only used a site to share pictures from special events and vacations, or to upload photos so that you can order prints, you've just scratched the surface. There are many different types of photo sites and a plethora of features, so it's worth looking around for one that lets you do all of the things you want to with your pictures.

Archiving Options

One very important use for online photo galleries provide is as a place to keep your images safe from any misfortunes that might befall your home digital photo collection. If your hard drive crashes, your house burns to the ground, or a burglar makes off with your computer, the photos you have uploaded to your online gallery will still be safe and sound in a server somewhere (or, ideally, two or more servers in different locations).

Of course, that will be a sure thing only if you make the right choices when you select an online gallery site to host your photos. Here are ten important factors to consider.

✔ **1. You'll need a broadband** Internet connection (usually cable or DSL for home users) to use an online gallery site as a practical archiving option.

✔ **2. Choose a site that stores your images at their full, original resolution.** The resolution of digital image files refers to their size, measured in pixels or megapixels (for example, 2400x1800 pixels or 4 megapixels; a megapixel is a million pixels). For archiving (and printing) purposes, you should always retain the full resolution of the original image. Some online services automatically reduce the resolution of photographs to take up less server space—a costly resource—and to make the images transfer faster to the Web. Avoid sites that perform such reductions, unless you only want to use them for sharing, and not to back up your photos. It's fine if a site creates and displays lower resolution versions of your pictures, as long as it also keeps the original-size photos.

✔ **3. For backup purposes,** it's best to use a site that makes it easy to download your full-resolution photos back to your computer or to purchase an archive disc containing them. Look for the ability to download batches of full-resolution files, not just single images, and make sure the price of archive discs seems reasonable.

✔ **4. On many sites, you must pay** for a level of service that archives full-resolution photos for the long term (at least a year). Some sites offer a choice between the paid subscription option and a free gallery that discards photos after a specified period and does not retain full-resolution files.

✔ **5. Many sites can host short videos** as well as photos. If you want to use a site as an online backup for your videos, make sure you'll be able to download them again or receive them on disc if needed.

✔ **6. Like all businesses,** photo gallery and sharing sites sometimes go under. This isn't always a disaster. In the past, major gallery sites that have shut down have given their members ample time and tools to transfer image collections to other sites. Look for a site that has been around for a while, looks well established, and advertises the security of its image archives. Avoid using a shaky operation that might take your photos down with it if it goes.

✔ **7. Gallery sites almost always impose a limit on the size** of each file you upload, although the limit is usually quite high. If you use a very high resolution camera or have large files of scanned images, check the limit before you sign up for a site. File size limits are usually more relevant to videos, restricting them to short clips.

✔ **8. All of the sites listed in this book are available in English,** but some can be used in other languages. Common alternatives include a variety of European languages, Russian, and East Asian languages. And yes, there is an option for all you Esperanto speakers out there. Just go to ipernity in Esperanto and click "Helpo" to find out all about it.

✔ **9. Many gallery sites only accept JPEGs.** However, there are a few that will let you upload TIFFs or other image file formats to your gallery. Photographers who want to archive RAW files online can use SmugMug's SmugVault or a professional media archive service.

✔ **10. Many of the online backup and sync services** and sites listed on page 42 include photo gallery tools. Whether you just need to get your photos online or you want to take advantage of their options for hosting other types of media and Web content, give those sites a look in addition to the top ten sites listed on page 111.

Ten Gallery and Community Tools to Look For

Online galleries and photo community sites offer oodles of tools for working with your photos online. These ten features can help you keep your images organized, share them with others, and move them between the collection of pictures you keep online and the one on your desktop. Consider which ones are most important to you when you're picking a site for your online collection.

✔ **1. Appealing gallery styles and structures.** When you're choosing a site, check out some sample galleries and find out the options for customizing yours. If the way galleries are organized doesn't make intuitive sense to you after a few minutes, or if you think they're just plain ugly, there's probably another site you'd be more comfortable with.

✔ **2. Flexible sorting options.** Look for a site that lets you sort your photos in a variety of ways—by date, name, caption, descriptive tag, or other criteria that can help you group your photographs into albums. This will make it easier to sort your shots in a way that makes sense to you.

✔ **3. Metadata support.** Check how the site allows you to use metadata—tags, keywords, captions, geographical information, and other descriptive information about your photos. Metadata can help you sort your online photo collection and group it into albums, give others a way to search for your pictures and understand their context better.

✔ **4. Export tools.** If you use a social network, have a blog or Web site, or use online auction and sale sites, look for an online photo service that provides tools for you to export copies of your images from your online gallery to other online destinations. This will allow you to upload photos just once to your gallery and use it as a central online repository from which to distribute images as you wish to other sites.

✔ **5. Mobile access.** Some online galleries can be viewed on a cell phone with data service and a mobile Internet browser. Having your whole online gallery accessible on your phone is a lot better than taking up your phone's memory with stored shots. Look for a site that provides a special mobile Web address or instructions for mobile viewing.

✔ **6. Software plug-ins.** Some sites provide software plug-ins that allow you to upload your photos directly from your image editing software. When you download and install the plug-in, the upload option appears on a menu or button in your software.

✔ **7. Sharing and privacy tools.** All sites give you at least some basic choices about whether your images can be seen by the public or are for designated eyes only. But you might want to share certain photos with friends and other photos with family. Think about whom you want to share your photos with and how, then look for a site that gives you the tools to do it. Some sites can also alert your contacts automatically when you add new images, let you e-mail them about updates directly from the site, or give you a platform for chatting with friends while viewing photos.

✔ **8. Photo editing tools.** Many people touch up their photos before they upload them to an online gallery, but if you use an automatic uploading tool to send them straight from your camera, camera phone, or hard drive, you might not get the chance. Look for

0 Zero Minute Gallery Uploads

Phanfare provides a desktop uploading tool that you can set to automatically update your Web albums. You download and install Phanfare's software on your Internet-connected computer, designate a folder to synchronize with your online galleries on the Phanfare site, and you're done. Whenever you add new photos to the designated folder, the software will automatically upload them to your Phanfare gallery, without you lifting a finger.

Some of the online backup and sync options described on page 42 in Chapter 1 offer this type of automatic photo gallery updating as well.

a site that provides effective tools for correcting your photos, including one-click image enhancement, brightness and color adjustments, and red-eye correction.

✔ **9. Printing services.** If you like to share your photos by creating photo albums or making prints for friends, make sure the site you select offers high-quality print services.

✔ **10. Backprinting.** To help organize prints after you've received them, some sites let you add text to the back of the prints. You might add dates, names of people and places, or the name of the photographer. Adding a message or caption to the back of a print can also be useful if you're making prints for others.

Online Resources ▸▸

Top Ten Gallery Sites for Photo Archiving
These sites let you upload a large quantity of photographs, store the images at their original resolution, and make your archive of full-resolution photos available to you on disc or in bulk downloads. They also offer long-term storage options.

Flickr www.flickr.com
Fotki www.fotki.com
ImageEvent imageevent.com
Kodak Gallery www.kodakgallery.com
Phanfare www.phanfare.com
Shutterfly www.shutterfly.com
SmugMug www.smugmug.com
Snapfish www.snapfish.com
Swiss Picture Bank www.swisspicturebank.com
Winkflash www.winkflash.com

More Online Gallery and Photo Community Sites
These sites offer a huge variety of styles, tools, and opportunities to share photos with people you know and people you don't. For example, serious photographers can join communities to discuss and critique their work on sites like Photo.net, JPG Magazine, and Pbase. (Or, for that matter, Flickr, listed in the top ten archiving sites above.) Frequent travelers will find plenty of travel-oriented options here, and online socializers can make themselves at home in sites that are set up more like social networks than typical photo galleries.

Adobe Photoshop Express www.adobe.com/products/photoshopexpress
AOL Pictures pictures.aol.com
dotPhoto www.dotphoto.com
DropShots www.dropshots.com
FotoTime www.fototime.com
Fujifilm.net www.fujifilm.net
GEOsnapper www.geosnapper.com
Google Picasa Web Albums picasaweb.google.com
ipernity www.ipernity.com
Jalbum jalbum.net
JPG Magazine www.jpgmag.com
KoffeePhoto www.koffeephoto.com
locr www.locr.com
Nikon my Picturetown www.mypicturetown.com
Panoramio www.panoramio.com
PBase www.pbase.com
Photo.net photo.net
Photobucket photobucket.com
Photocheap www.photocheap.biz
Photomax www.photomax.com
PhotoWorks www.photoworks.com
Picturetrail www.picturetrail.com
Piczo www.piczo.com
pikeo www.pikeo.com
Pixagogo www.pixagogo.com
Plazes www.plazes.com
Printroom www.printroom.com
Sacko www.sacko.com/default.aspx
TripTracker triptracker.net
More on page 184

Distributing Your Photos Online

These days many of us post our photographs in more than one place online. You might be part of a social network or have a blog or Web site where you show photos to friends, family, or the public. Posting photos can also serve a practical purpose when you're selling something online, and if you use forums to discuss your interests with others, you may occasionally post images to illustrate a point.

If you have multiple online destinations for your photos, you probably know how to post them directly to your favorite sites. Each site you use provides its own uploading tools and instructions for adding photos from your computer. But there are ways to streamline the process of distributing your photos to multiple destinations and to keep your online photo collection organized. You can use your online gallery or media archive as a central repository, and put automatic distribution tools to work.

When you select an online gallery or media archive, look for tools that let you distribute your images to other sites. This distribution is like turning on a TV at the destination sites, on which viewers can see images that are stored in your central repository. You're not actually sending copies of your photos all over the Web, you're just broadcasting them. Different host sites describe their tools for distributing photos in different ways. Some talk about "sharing" and "embedding" photos in other sites, while others give you the option to use a "widget" or simply to "blog this," as Flickr puts it.

The way these tools work is that the host site—the one to which you upload your photos—automatically generates a string of HTML (HyperText Markup Language). When that string is pasted into a destination site, your photos appear there. Some host sites automate the process of sending the HTML string to the destination site and pasting it in, while others require you to copy and paste it manually.

Some sites, such as Photobucket, are designed to be used as a central repository for images that will be distributed to numerous destinations. They may provide strings of hypertext that you can paste into e-mail, an instant message, or a post to a forum.

Other sites offer automatic tools for exporting photos to just one or a few online destinations. As long as the host site shows you the HTML that it generates, however, you should be able to copy and paste that string into a variety of destinations. When you navigate to the destination site, just log in and go to an editable version of the page where you want your image to go. Then paste the HTML into the area where the site lets you enter text and code. For example, here's how you would paste hypertext into your profile if you're a member of the MySpace social network:

-Log in to your account.
-Click on Edit Profile on your MySpace home page.
-Paste the HTML into the About Me box that appears.
-Save your changes.

About Me:

```
<img src="http://farm3.static.flickr.com/227/2157799_x2e3x5xt6_o.jpg" width="188"
height="216" alt="self_portrait" />
```

Preview Section | Preview Profile

The image you distributed will appear in your MySpace profile under About Me. The nice thing about this type of distribution is that it often allows you to broadcast not just one but many photos from your central repository. Many sites provide tools for creating slideshows that can be distributed, so that the slideshow plays on the destination site. SugarSync (see page 181) generates HTML that broadcasts all of the images in a designated album, adding new photos to the continuous slideshow as they're uploaded to the SugarSync gallery.

0 Zero Minute Photo Distribution

You can completely automate distribution of your images—from your camera to multiple online destinations—if you choose sites and downloading systems that give you the right tools. You can set up fully automatic distribution in three steps:

✔ **1. Use a wireless downloading system** that automatically transfers your photos from your camera to your computer. These can include an Eye-Fi SD card (see page 95), a WiFi digital camera (page 95), or Kodak Picture Upload Technology (page 99) if you're using a camera phone.

✔ **2. Use an online gallery** or media archive site that provides an automatic photo uploader. Good examples include Phanfare (page 110) and SugarSync (page 181). Once you install the site's desktop software and designate a folder from which to upload photos, your images will be automatically uploaded to your online gallery when your computer is turned on.

✔ **3. Use an online gallery** or media archive site that gives you tools for automatically distributing newly uploaded images to blogs, social networks, and other designated destinations.

Once you've set up your wireless downloading system to automatically move your photos from your camera to your computer, set your automatic uploader to move them to your online gallery, and set your online gallery to distribute them, they will travel from your camera to all of your online destinations without you having to do a thing. Just remember that this system is in place the next time you snap a bunch of shots of your friends skinny-dipping.

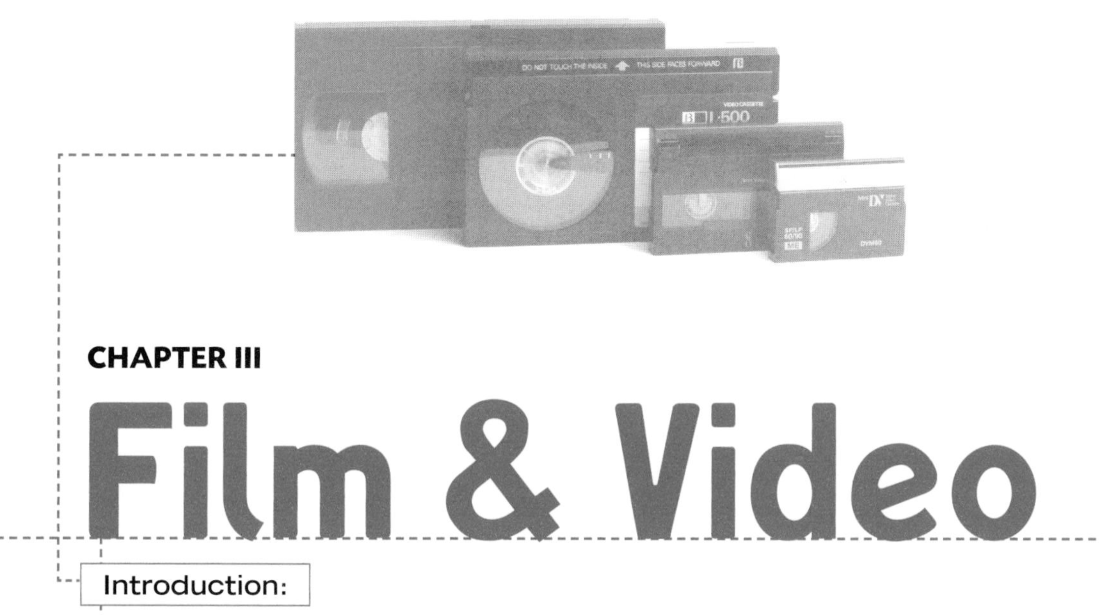

CHAPTER III

Film & Video

Introduction:

It's easy these days to take thousands of photos with a digital camera, and we all treasure our photo collections. But there are some things that a still shot just can't capture. A child's first steps, your mother's comic timing, or the first dance at your daughter's wedding—these kinds of moments are best preserved in full motion with sound.

If you and your family have been recording such moments over the years, you have inevitably amassed a collection of outdated film and video media, and there's a good chance you haven't watched your home movies in years. Transferring your movies to a digital format can let you watch them again and share them in ways that had not even been invented when they were first shot.

In this chapter, you'll learn how to digitize analog footage on film and video tape yourself, as well as what to look for if you decide to send it to a professional service. You'll also find ways to organize, preserve, and archive your digital video collection and to share your masterworks online.

Digitizing Analog Video Tapes

The Goal:

Create digital video files from your analog video cassettes so that you can integrate them into your digital video collection and easily display and share them with media player software and devices.

What You Need:

A device that can play the video tapes and output the analog video, a device that can convert the analog signal to digital, and a device that can record the digital video. As described here, there are numerous setups that include these three elements.

Time Required:

The recorded time of your videos, plus about fifteen minutes per video to name files and add menus and chapter markers.

There are even more good reasons to digitize the home movies you have on video tape than there are to digitize other kinds of analog media. Of course, there are the usual advantages of being able to play and share your videos with current software and devices. But another reason is even more compelling: consumer-grade video tapes can deteriorate relatively quickly. If you have video tapes that are ten or twenty years old, the images may already have begun to deteriorate—especially if the tapes weren't of the highest quality to start with, or weren't stored properly. If your tapes are more than twenty years old, you're starting to push your luck.

If you have only a few tapes and no longer own any devices that can play them, the most economical way of digitizing them might be to use a transfer service. The cost of having video tapes digitized starts at about $10 per tape. (Read more about transfer services on page 124 .) If you want to digitize more than a few tapes, the cheapest option will probably be buying any necessary equipment that you don't own and doing it yourself. The process isn't difficult and you can get good results doing it yourself, although it does require some time.

The first thing you need to know in order to digitize your tapes is whether they're analog video tapes. Both analog video tapes and digital video tapes just look like tapes, and people often talk about "digitizing" tapes that are already digital, when they simply mean transferring the footage to a computer. To clear up the confusion, here are the basic facts: Analog and digital cassettes both record video data on a magnetic tape. However, the way the data is recorded is different. Analog tapes record an analog signal, while digital tapes store video data as zeros and ones, just like any digital medium. As a result, analog video has to be converted into a digital format before it can be saved on a computer hard drive or an optical disc. Digital video, on the other hand, can be transferred directly from any camcorder or deck that can play the cassette, without any analog-to-digital conversion process.

Here's a rundown of common analog and digital video cassette types, to help you sort out your collection:

ANALOG CASSETTE TYPES	DIGITAL CASSETTE TYPES
VHS	MiniDV
VHS-C	Digital8
S-VHS	MicroMV
Betamax	DVC Pro
Video8 (8mm)	DVCAM
U-matic	
Hi8	

Digitizing Setups

The equipment you'll need to digitize your analog tapes depends on the type of video cassette you have, but in every case it must link three basic elements: a device that can play the cassette, an analog-to-digital converter, and a device that can record the digital output as the video is played and converted. You also need to select video editing software. (The Resources section on page 133 lists video editing software options.) If you buy a video conversion device, it will probably come with software.

When you connect video devices with cables, you may sometimes have a choice between connecting via composite video (RCA) jacks or S-Video. RCA video connectors are yellow and have a single large metal pin, while S-Video connectors have four small round pins and a larger rectangular pin. Always use S-Video if it's an option. It produces a higher-quality signal. Don't forget to connect your audio jacks as well when you're outputting video from a camcorder, VCR, or deck. There will be two audio connectors on your A/V cable, one red and one white. Whether you use the yellow RCA video connector on your A/V cable or a separate S-video cable, you will use the same A/V-cable audio connectors for sound.

Here are some common setups:

- **Dual VCR/DVD deck.** This is a deck that has a slot for VHS cassettes and an optical drive for DVDs. It has a built-in analog-to-digital converter, and can record from VHS to DVD. A dual deck provides one of the easiest ways to digitize your VHS tapes, since all you have to do is hit Record on the DVD side and press Play on the tape side. Make sure you buy a deck that can record DVDs, though. Many dual decks can play DVDs but not record them.

- **Digital8 camcorder to computer.** If you have a Digital8 camcorder and analog Video8 or Hi8 cassettes, you may be able to use the camcorder to digitize the analog video. Some Digital8 models have built-in analog-to-digital converters, which allow them to play analog tapes. Check the camcorder manual or specs to find out if yours has this capability. If so, put the analog tape in the camcorder, then connect the camcorder to your computer with a FireWire (IEEE 1394) cable. Your computer must have a FireWire port for this to work.

- **Camcorder to converter to computer.** If you still have an analog camcorder that can play your tapes, you can connect it to a device that will convert the analog signal to a digital one, then connect that device to your computer. Various companies sell conversion devices that you connect to your analog camcorder with an A/V or S-Video cable and to your computer via a USB cable. If you're comfortable opening up your computer and installing a card in a PCI or PCI-e slot, another option is to buy an internal adapter for your computer that has RCA and S-Video inputs and an on-board analog-to-digital converter. Adapters like these are included in ATI's TV Wonder product line.

- **VCR or deck to converter to computer.** This setup works the same way as the camcorder to converter to computer setup, except that the camcorder is replaced by a VCR or other video tape deck.

- **Camcorder or VCR to disc burner.** Some standalone DVD burners have analog inputs and a built-in analog-to-digital converter, allowing you to connect an analog camcorder or VCR to the burner with RCA or S-Video connectors and record directly to a DVD. Look for a model that has a built-in screen so that you can see the video play while it's recording. If you want to use a disc burner to digitize your videos, but also wish to edit them, first burn the video to a DVD-RW or DVD+RW rewritable disc, then put the disc in your computer and import the video into your editing software. After you make the desired adjustments, burn the final edited video to a DVD-R, DVD+R, or Blu-ray disc if you want to archive it on disc. You can reuse the rewriteable disc to transfer more videos.

The Analog Video Transfer Process in Ten Steps

Once you've decided on a transfer setup and hooked everything up, here's what you need to do. The details of this process vary somewhat depending on the setup and software you use.

✔ **1. Clean your player.** If you're using an old VCR or camcorder to play your analog tapes, buy a head-cleaning cassette and give the device a cleaning before you start digitizing. That way, you will be sure that your deck is clean and working properly, so it won't eat your tapes. (If you're using a dual deck or recording directly to DVD on a standalone burner, you'll simply follow the manufacturer's recording instructions from this point on. If you decide to transfer the video to your computer, you can go to step 6.)

✔ **2. Check your PC's audio input.** If you're connecting a conversion device to a Windows-based computer via USB, go to Control Panel/Sound and make sure the USB audio input is selected.

✔ **3. Open your video software and check the audio input.** Check the preferences or options and make sure that the correct audio source is selected.

✔ **4. Check your import settings.** Make sure your software is set to import the video in the format you want. Read more about video formats on page 122.

✔ **5. Digitize the video.** Hit Import or Capture in your software and play the video on your player or camcorder. Try not to leave a long blank space at the beginning of the clip, and stop recording as soon as it's done.

✔ **6. Adjust the brightness and color.** Your software should provide tools for enhancing the brightness and color of your video. Select the whole clip, then apply the appropriate tools. Take a look at some of the darkest and brightest frames to make sure the exposure looks right.

✔ **7. Add transitions and trim clips.** Fade-outs and fade-ins allow you to move from scene to scene and trim off any blanks or unintended recordings—like when the videographer put the camera down to play ball and recorded the grass growing for ten minutes.

✔ **8. Insert chapter markers.** These will allow you to search for and jump to scenes by viewing a chapter image index. If your software doesn't create chapter markers automatically, then insert them manually. You can do this by creating a marker at the beginning of each transition to a new scene, or by simply adding a marker at regular intervals—say, every five minutes.

✔ **9. Name the video and create a menu.** Give your video a clear descriptive name. If you're planning to burn it to a DVD, use your software to create a menu for navigating to chapters.

✔ **10. Export, or save, the video in the format of your choice.** Save the files to your Video folder so that your media player will find them easily.

Video Conversion Resources ▸▸

Dual VCR/DVD Deck Makers

JVC www.jvc.com
Panasonic www.panasonic.com
Philips www.philips.com
Samsung www.samsung.com
Sony www.sony.com
Toshiba www.toshiba.com
Zenith www.zenith.com

Analog-to-Digital Conversion Device Makers

ADS Tech www.adstech.com
AMD www.amd.com
Belkin www.belkin.com
Pinnacle www.pinnaclesys.com
Plextor www.plextor.com
StarTech www.startech.com

Digitizing Films

The Goal:

Create digital video files from your home movies on film.

What You Need:

A film projector that is compatible with your type of film and offers variable speed control; a digital camcorder with manual focus and shutter speed controls; a projection surface; and cleaning solution with a soft cloth or film cleaning pads. A television for monitoring the image while digitizing, white cotton gloves, and video editing software are also desirable.

Time Required:

The recorded time of your film, plus time to set up and adjust your equipment and to transfer your digitized video to a computer and do some basic editing.

Transferring your old home movies from a film format such as 8mm, Super 8, or 16mm isn't quick or simple, but if you have access to the right equipment and you're willing to devote a substantial amount of time to the project, you can get decent results. If you want really good results, you're probably better off using a professional film transfer service. That's also true if your film is damaged or broken, and you don't have the tools or expertise needed to splice it back together and repair it. Read more about those on page 124.

Digitizing Films in Ten Steps

✔ **1. Clean and lubricate the film and projector.** It's very important to clean and lubricate your film before projecting it with a special film cleaning solution. Wear a pair of white cotton gloves so that you don't leave fingerprints or oil smudges on it. You'll need to move the film from the reel it's on to a takeup reel as you clean it. You can do this by mounting the reels on your projector and manually winding them. This method is a little clumsy. If you have a lot of film, you may want to buy a pair of rewinds. Whichever system you use, set up the film, then pour some cleaning solution on a soft cloth, such as a microfiber cloth, or on a cleaning pad. Fold the cloth or pad over the end of the film and start winding the film onto the empty reel, pulling the film through the cleaning solution as you wind it. Make sure you give the cleaning solution time to evaporate before the film is wound.

✔ **2. Set up a projection surface**. In order to retain the quality of your film image when you record it, you need to project it at a small size and prevent light from striking the projected image. You can project it on a white wall or posterboard, or you can make a homemade transfer box that shields the screen from light. Paint the inside of a rigid cardboard box

Film Cleaning Solution Makers ▸▸
ECCO
Edwal
FilmRenew
VitaFilm
Xekote
prosites-clementsx.homestead.com

with flat black paint, then line the bottom with a piece of white paper. When you turn the box on its side, you will have a little theater (without seats) with a white projection screen at the back. The "screen" doesn't need to be any larger than about a foot wide. You can also buy used commercially made transfer boxes online. They vary in quality. Some use mirror systems so that you can put your camcorder on one side of the box and your projector on the other side. This can make the equipment easier to align.

✔ **3. Set up the projector.** Put your film reel on the projector, and place the projector in front of your projection surface. Turn on the light and focus the projected image.

✔ **4. Set up the camcorder.** Put the camcorder on a tripod or surface near the projector, and position the lens so that it can capture the image that will be projected. Switch the focus to manual focus. Turn off any image stabilization functions. Familiarize yourself with the camcorder's exposure controls; you'll need to adjust them while recording. If you're digitizing a film with sound, connect your projector's audio output to your camcorder's audio input with a cable if possible.

✔ **5. Adjust your framing and shooting angle.** Start projecting the film so that the frame size is about 8 inches wide or smaller. Adjust the angle of the screen and camcorder to get a good image. Adjust ambient lighting to eliminate hotspots on the screen. Zoom in on the screen with the camcorder, and focus manually on the image until it looks sharp. Then zoom out until the film image fills the camcorder viewfinder. Look at the image on the camcorder's LCD to see if it is in focus from edge to edge, and adjust the angle of the camcorder or screen until it is. To make sure that the focus is sharp, connect your camcorder to a TV or computer monitor and look at the viewfinder image there.

✔ **6. Adjust the film and camcorder speeds.** To avoid a flickering image—the result of a difference in the rate at which the projector is displaying frames of film from the one at which the camcorder is capturing them—synchronize the shutter speed of your camcorder and the variable speed control of your projector.

✔ **7. Record the movie.** Rewind your film to the beginning, press Record on your camcorder, and start projecting the film. Monitor the image on your connected TV or monitor.

✔ **8. Transfer the movie to your computer.** Connect your camcorder to your computer in the same way you usually do, open your video editing software, and import your digital video.

✔ **9. Adjust the brightness and color.** Select the whole clip when doing this. Take a look at some of the darkest and brightest frames to make sure the exposure looks right.

✔ **10. Export, or save, the video in the format of your choice.** Save the files to your Video folder so that your media player will find them easily.

Understanding Digital Video Formats

The digital video format you use for recording depends on the formats that are available from your camcorder, digital camera, or camera phone. But when you transfer videos to a computer or burn them to disc, you have the option to convert them to a different format. Knowing something about how the various video formats compare helps you retain the original quality of your videos, as well as the ability to edit them. It also gives you a basis for making decisions about tradeoffs between storage space and quality.

Here are two rules of thumb for choosing a video format to save and archive your footage:

- **Don't down-convert your original.** If you captured your video in a high-quality format, don't convert it to a low-quality format for archiving. You can always make lower-quality copies as needed for playback on devices or online.

- **Don't lose high-quality editing capability.** Formats that use interframe compression are generally not as well suited for editing as formats that use intraframe compression. Interframe compression works by discarding some visual information that is repeated in sequential frames to make the file smaller. When you open or play the file, your software fills in the missing information in each frame by looking at similar frames and deducing what is missing. If your original file uses an intraframe compression format such as DV, don't convert it to an interframe format before you edit it. By the same token, if you're having film transferred to a digital file that you want to edit heavily, it's best to have it transferred to an intraframe compression format.

Figuring out which file format you're dealing with can sometimes be confusing, due to the nature of video files. A digital video file is actually a package of files. On the outside, there's a media container. Inside it, there is a file that holds the visual information for the video and a file that holds the audio information for the sound. A video file's extension tells you the format of the container, but it doesn't necessarily tell you which video and audio formats are used by the files inside. For example, your file might end with .avi, but the video inside could be anything from a high-quality DV footage to a Motion JPEG clip captured by a snapshot camera. All that the .avi extension tells you is that the video is in an AVI media container. Be careful when you're converting video to a different format that you select the right video format, instead of selecting by media container type. Video formats are tied to resolution limitations. Those that store standard-definition video have a resolution limit of 720×480 pixels, while high-definition formats allow higher resolutions.

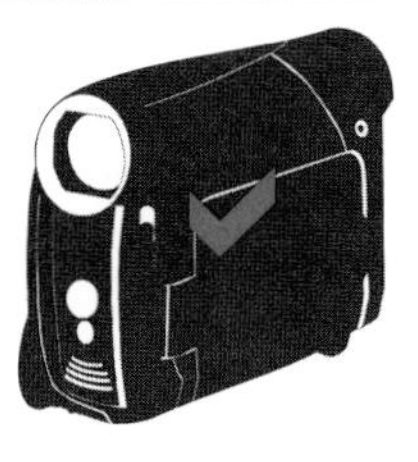

A Quick Overview of Popular Video Formats

The following are formats used by camcorders to save captured video. Video editing software can also convert digital video between these formats.

FORMAT	PHYSICAL CAPTURE MEDIUM USED BY CAMCORDER	COMPRESSION TYPE AND FILE SIZES	CHARACTERISTICS
DV	MiniDV cassette	Intraframe compression, large file size	Standard definition; good for editing and archiving
HDV	MiniDV cassette	Interframe compression, large file size	High definition
MPEG-2	DVD, hard drive	Interframe compression, moderate file size	Standard or high definition; good for archiving on DVD
AVCHD	DVD, flash memory, hard drive	Interframe compression, moderate file size	High definition; good for archiving on DVD

Digital cameras capture video clips in numerous formats, including MPEG-1, MPEG-4, DivX, H.264 (AVC), and Motion JPEG. These formats generally offer lower quality than the formats used by camcorders, but they are more compact, yielding smaller file sizes. The MPEG-4, DivX, and H.264 formats come closest to MPEG-2 quality, but they compress video to a smaller size. Motion JPEG offers decent quality, since it is essentially a series of JPEG still images that are compiled to create a video file. However, Motion JPEG also requires much more storage space than the other formats. That can limit the length of the clips you capture, unless you have a very high-capacity memory card.

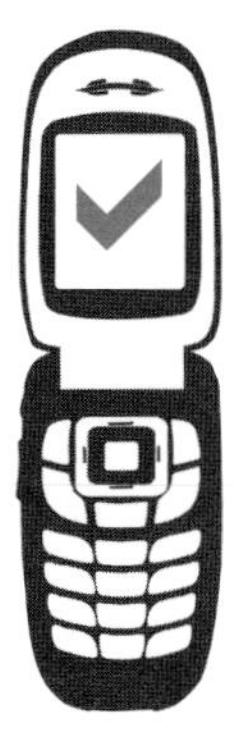

Camera phones usually use the 3GP format, which offers low quality but is very compact, so that it can be transmitted on a cell network. Some camera phones can also save video in the MPEG-4 format, which offers higher quality. MPEG-4 is a good option to look for if you use your camera phone to shoot video frequently and want to use it for purposes other than just sending it to your friends' cell phones. Don't convert your higher-quality video to any of these formats for archiving purposes. If you want to upload video or play it on a device, find out which format the site or device requires, then make a copy of your video in that format.

Using a Home Movie Transfer Service

The Goal:

Have your collection of films or analog video cassettes transferred to a digital format so that your home movies can be played with current devices, edited on a computer, shared online and on disc, and archived for safekeeping.

What You Need:

Money and a shipping box.

Time Required:

Much less working time than it would take to do your own transfers, especially in the case of film. (All you have to do is pick a transfer service, place an order, and pack up your home movies.)

If you have reels of home movies on film, and you're not a film-transfer buff, you should seriously consider sending them to the professionals for transfer to a digital format. There are numerous factors that can cause a film transfer to turn out badly, and it's highly unlikely that you own the kind of high-quality equipment that the better transfer services use. Film transfers aren't exactly cheap, but if you use a service with high-end equipment and experienced technicians, you will definitely be getting something of value for your money.

It's much easier to get good results digitizing analog video tapes yourself, but if that would require you to spend a substantial amount on equipment or you'd just rather avoid the hassle, you can send your old cassettes in for transfer, too. Take a look at the sections in this book on digitizing your film and analog video cassettes to get an idea of the equipment you'd need to buy to do it yourself, and compare the costs to the estimated expense of using a service. Time is also a factor when you transfer movies from film and analog cassettes, since the transfer happens at the speed of playback. If you have fifty hours of footage to transfer, you may decide that you'd rather pay someone else to tackle the job than spend fifty hours on the project, plus setup and editing time.

Sending in digital video cassettes for transfer to DVD doesn't usually make sense unless you are really short on time. You probably still own the devices that can play them and transfer them to a computer, and the process isn't difficult. Of course, if you have digital video cassettes in an unusual format, such as Sony's short-lived MicroMV, and you no longer own a device that can play them, it might be cheaper and easier to pay a service to transfer them to another format than to acquire the necessary equipment.

You can find numerous mail-in transfer services online, and there may also be options in your local area. The way that transfer services work is that you ship your reels or cassettes to the company, their contents are captured and saved as digital files, and then the company ships your reels and cassettes back to you, along with your newly digitized movies on disc, digital cassette, or hard drive. If you want your movies to show at an upcoming event, make sure you allow ample time. Services may need weeks or even more than a month to complete the process.

Ten Home Movie Transfer Service Features to Look for

✔ **1. High-quality file formats.** Many services offer to transfer analog movies to DVD in the MPEG-2 format, but that's not the best option for films in terms of quality. If you want to retain the highest image quality, choose a service that can send you your digitized films in a high-quality format such as an uncompressed or DV file on a hard drive or MiniDV cassette ▶. (Read more about video formats on page 122.) Expect the file sizes to be quite large, in the range of 12GB for an hour of video. Analog video cassettes often have lower image quality than films, so you probably won't lose much in transferring them to the MPEG-2 format on DVD. Don't have your film transferred to an analog cassette format such as VHS. You will lose a lot of quality and be stuck with a cassette that is itself outdated. The only exception might be if you want to give the VHS tape to someone who is elderly and not likely to upgrade to a digital player. In that case, get a digital master made for your own archive and then have a VHS tape copy made for them. If you do have your film transferred to DVD, one disc should hold no more than two hours of footage—more than that, and its quality is probably being compromised.

✔ **2. Editable files.** If you think you will want to edit your video or combine it with other material in a multimedia show, make sure you receive the digital files in a format that can be edited. The best choice is to receive a MiniDV cassette or hard drive containing AVI files of DV footage. (To use the MiniDV option, you will obviously have to have a MiniDV camcorder or deck available to get the video onto your computer for editing.) Most services that offer high-quality editable file types can also send you copies of the video on DVD, so that you have both the high-quality video for archiving and editing and the convenience of simply popping a DVD in your player to see the movies.

✔ **3. Frame-by-frame transfer.** Some services use systems that scan each frame of film separately instead of projecting the film at normal speed and recording the projection with a camcorder. If done properly with high-end equipment, this process can result in better image quality.

✔ **4. Full-frame capture.** Make sure the service you choose captures the full area of each frame of your film, without cutting off the edges of the image.

✔ **5. Soundtrack option.** Make sure the service you use to transfer from film can handle sound capture if your film has a soundtrack.

✔ **6. Included splicing and basic editing.** Repairs on damaged film usually cost extra, but look for a service that will include some basic splicing and editing to take out blank spots and repair simple film breaks.

✔ **7. Included menus, chapters, and disc labels.** If you order copies of your digitized video on DVDs, they should include menus and chapter markers so that the video is easy to identify and search. Find out if the service can also label your discs.

✔ **8. Color and brightness correction.** The transfer service should be able to enhance and balance colors and adjust exposure, so that the digital video has a vibrant but natural look. (However, if your original is very faded, there's only so much that can be done to restore it.) Find out if these improvements will cost extra. Also look for noise reduction for video from cassettes, to make them look less fuzzy. Adding special effects such as transition effects between scenes often costs extra.

✔ **9. Foreign cassette format support.** If you have video cassettes that were made outside of the U.S., they may use a recording standard that's different from the American NTSC standard. Make sure that the service you choose can transfer from PAL, SECAM, or another foreign video format if necessary.

✔ **10. Secure shipping options.** If you're worried about shipping losses, break up your reels or tapes into several packages and send them separately. The chances of one package being lost by a reputable transfer service and shipping company are extremely low. The chances of more than one package from the same customer being lost are virtually nil.

▸▸ Transfer Cost Estimate Worksheet

This worksheet will help you compare the prices and services of various video and film transfer companies. Fill out a separate copy of the worksheet for each type of media that has a different base transfer price (for example, 8mm film, VHS tapes, and so on). Using this worksheet won't give you the precise amount your transfer will cost, since it does not include shipping costs and you may be estimating the recorded length of your film or videos. It is designed to help you get a ballpark idea of what having your home movies transferred will cost, and to find a service that fits into your budget.

COMPANY NAME	TOTAL FOOT LENGTH OF FILMS / # OF CASSETTES*	OUTPUT MEDIA	BASE FEE PER FOOT/ CASSETTE	FEES FOR OTHER SERVICES	FEE FOR ADDITIONAL COPIES	NUMBER OF COPIES DESIRED	TOTAL COST
Example company	400 feet	DVD	15¢ per foot	$50 flat repair and splicing fee	$10	3	
Subtotal			(400 x .15=) $60	$50		(10 x 3=) $30	$140
Subtotal							
Subtotal							
Subtotal							
Subtotal							
Subtotal							
Subtotal							

*Enter the number of video cassettes or feet of film from your Media Collection Inventory (page 11).

Organizing Your Digital Videos

The Goal:

Create an organizational structure for your digital video files, put your existing files in it, and choose tools for managing your movies.

What You Need:

Your computer and any hard drives, devices, discs, or memory cards where you have digital video stored; video editing or player software.

Time Required:

About ten minutes to set up an organizational system for your home videos. Additional time to transfer your existing videos into it and select video management tools.

Most if us have a few categories of digital video stored on hard drives, discs, and cassettes. First, there are home videos and other amateur productions that we take ourselves with a camcorder. Then there are those little clips that we capture with our digital still cameras or camera phones. And finally, there are commercial videos, which it's becoming easier to accumulate, now that online video services are expanding the selections that they offer for download and improving the technology for downloading them.

When you organize your videos, it's best to address these categories separately. Don't try to combine your home video collection with your commercial video collection. They have different uses, and they are most easily managed in different ways. From an organizational perspective, treat your home videos like your photo collection and your commercial videos like your music collection, and then decide whether to keep your camera and camera phone clips with your photos or your home videos.

The following sections offer closer looks at how to get each segment of your video collection in order.

Organizing Your Home Videos

In your Videos or My Videos folder, create a folder system for your home videos that is similar to the one you created for your photos (page 76). It makes sense to organize them chronologically by year. Within each year, unless you are a real video diarist, it usually works best to use subfolders named for events and projects, not monthly folders. Make sure that the date and time on your digital camcorder are set correctly, so that your video files have that information attached to them for searching and sorting.

Create your own variation on the suggested folder structure ▼ to meet your specific needs. To set up your folders, use the Explorer file browser (this is your desktop browser, not Internet Explorer) on a Windows system. On a Mac, use one of the file browsers listed on page 76. First, open your Videos folder and create new subfolders going back one year and forward one year. Once you've created the folders, use your file browser to search your computer for video files, sort them by date if possible, then move them into the appropriate folders and make sure that they have meaningful names. If you can't identify the files, you may have to open them with your video player software to see what they are before you rename them.

Video files are large, so you may not be able to store footage from previous years on your computer. In that case, keep them in your digital video archive. You can organize videos from previous years on your computer and then move the annual folders to a different storage device to make room on your system. Read more about archiving videos on page 132.

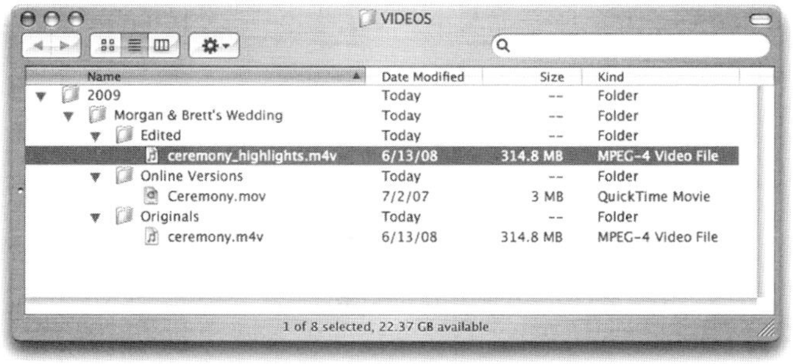

Originals. Save unedited video from your camcorder to this folder. Also save footage transferred to a digital format from analog cassettes or film here.

Edited. When you use software to edit your videos, save the edited clips as copies here, and leave original footage in Originals as an archived version. If you or another person want to use new software or techniques to edit your video in the future, it will be advantageous to start with the full original file.

Online Versions. You may create copies of your videos in a compact format to stream or share online. You can use this separate folder to store those small files and make them easy to find when you're uploading or setting up an automatic uploader for an online gallery site.

Ten Key Home Video Management Tools

With your home videos organized in folders, you can search and manage them through media player software and work with them in video editing software. If you have a lot of video clips, you may want to use dedicated video cataloging software such as iDive for the Mac. Once you import your videos into the software from your hard drive or an attached device, or use an editing program's capture tools to acquire footage from a MiniDV or HDV camcorder, the software keeps track of your video files with its built-in database cataloging system.

Here are some features to help you keep videos organized as you work with the files and watch them. Look for them in the software you select, and make sure it includes those that you need most.

✔ **1. Automatic indexing.** This function automatically detects scene changes in your video and creates a scene index of thumbnail images. It can usually be run as you import video into your editing software. The way that different programs do this varies and can be more or less effective. The way the function works with different types of video varies as well. This feature is especially useful with DV footage from tape. Look for software that can do a fast scan of your DV video and automatically index it.

✔ **2. Chapter marking.** This allows you to divide your video into chapters and mark the beginning of each chapter. You should be able to automatically generate chapter markers at regular intervals and to adjust the placement of markers in your video on a timeline. When you burn the final video to a disc, the chapters appear in the main menu so that you can select scenes easily.

✔ **3. Tagging and search tools.** Some software lets you attach keywords to video files and tag them with ratings or categories such as People, Events, or a custom category. This allows you to search and sort clips within the program.

✔ **4. Automatic file naming.** Look for a function that automatically names your files, using a scheme that you have specified, as the video is captured into the software from a MiniDV or HDV camcorder. Some programs can automatically rename video files with the date when they were captured when you import them.

✔ **5. Clip splitting and merging.** This allows you to split long video clips into separate files or to merge files together. Look for software that gives you intuitive ways to do this.

✔ **6. Sync with media receivers and portable devices.** If you have a portable video player or an A/V receiver connected to a home network, look for media player software that can sync video files with the device or stream them to it.

✔ **7. Web export tools.** If you want to share your videos online, look for editing software that gives you simple tools for creating versions of your videos in a format that can be played online. Some software makes it easy to generate and upload online videos directly from the program.

✔ **8. Clip list export.** Some programs can create a list of your video clips and export it as a text, spreadsheet, or Web file.

✔ **9. Titling tools.** Look for easy-to-use titling tools so that you can include text in your videos. This lets you add information and credits to the beginning or end of your videos, or even to add commentary or subtitles in other places.

✔ **10. Annotation and logging tools.** These features are useful for large video projects, and allow you to attach text notes to your video clips so that you can keep track of different segments while you're assembling and editing a complex video piece.

Organizing Your Camera and Camera Phone Videos

For most people, keeping video clips captured by a digital still camera or camera phone in a separate folder doesn't make sense. Keeping them with other videos or with photos that they are related to generally works well and keeps things simple.

If you take video clips and photos with the same device, you'll probably want to keep the two together so that you can search for and find all of your materials from a particular occasion or date in one place. On the other hand, if you take a lot of camera or camera phone video clips and want to edit them or include them in multimedia projects, it might make more sense for you to keep them with your video files in thematically organized folders instead. If this is the case, you'll need to manually move your video files from your photo folders after you download them, since they will likely end up wherever you have set your downloading software to put your photo files.

Organizing Your Commercial Videos

Keeping the videos you purchase from commercial sources organized usually isn't a problem. Just use the software player provided by the service through which you downloaded the videos to search, sort, and manage them. For example, if you buy videos from Apple's iTunes Store or Amazon Unbox, manage them with the iTunes or Unbox software.

When you purchase commercial videos, they should come with complete information about the movie or show, so you won't have to worry about unidentifiable files accumulating on your system. Just keep your commercial video files in the folder where they were downloaded by your video service, and the player should have no problem locating and managing them.

If you run out of space on your computer's hard drive, you can move the files to an NAS device or external hard drive and then update the location of the videos in your player software's preferences so that it can find them.

Archiving and Sharing Digital Videos

Digital video files are big. And the higher their quality, the bigger they are. A single hour-long MiniDV cassette holds about 12GB of data. It also takes time to transfer cassette-based formats such as DV and HDV to a computer, because they must be captured in real time by video editing software from a cassette in a connected camcorder. For these reasons, you may find that storing your videos on your computer and backing them up to your usual backup drive aren't practical options. Uploading high-quality videos to online storage for safekeeping isn't a practical option, either. But if it was important to shoot the videos in the first place, it's important to store them securely, and to make it easy to find and watch the footage you're looking for. You should avoid down-converting your video files to lower-quality, more compressed formats just to save storage space. Here are some ways that you can manage large home video collections, and some tips on keeping them organized.

Options for Backing Up and Archiving Digital Video Cassettes

Use video editing software to capture your video from a connected camcorder, then save it on a high-capacity external hard drive or NAS. Store your original cassettes in a separate location.

Use a video editing program that has a function for recording video back to a cassette. Capture your video with the software, then record it to a second cassette, and store the duplicate cassette in a different location.

Use your video editing software to capture the video from your connected camcorder, then save it to DVDs or a Blu-ray disc as a data file. This preserves the full quality and editing flexibility of the video files, but requires more space than burning the video in MPEG-2 format. You'll need three single-layer DVDs to store a single MiniDV cassette worth of video, so you may want to use dual-layer DVDs or Blu-ray discs to reduce the number of discs you have to store.

Output your video from a camcorder to a standalone disc burner or use a video editing program with a function that can capture the video from a connected camcorder and burn it directly to a disc in MPEG-2 format. Store the discs and the cassettes in different locations. You will lose the ability to edit a high-quality DV original if something happens to the cassettes and all you have left is the discs, but at least you won't lose your videos completely.

Video Organizing Resources

Media Players with Organizing Functions
Adobe Media Player get.adobe.com/amp
Apple iTunes www.apple.com
Microsoft Windows Media Player www.microsoft.com
Nullsoft Winamp www.winamp.com
RealNetworks RealPlayer www.real.com

Video Editing & Organizing Software
For casual videographers
Adobe Premiere Elements www.adobe.com
Apple Final Cut Express www.apple.com
Apple iMovie www.apple.com
Arcsoft VideoImpression www.arcsoft.com
CyberLink PowerDirector www.cyberlink.com
Microsoft Windows Movie Maker www.microsoft.com
Nero Vision www.nero.com
Pinnacle Studio and Studio Plus www.pinnaclesys.com
Roxio Easy Media Creator www.roxio.com
Sony Vegas Movie Studio www.sonycreativesoftware.com
Ulead VideoStudio www.ulead.com
For serious videographers
Adobe Premiere Pro www.adobe.com
Apple Final Cut Pro www.apple.com
Pinnacle Studio Ultimate www.pinnaclesys.com
Sony Vegas Pro www.sony.com

Video Management Tools
Aquafadas iDive www.aquafadas.com
DVdate paul.glagla.free.fr/index_en.htm
ScenalyzerLive www.scenalyzer.com

Cataloging Commercial Movies
Collectorz Movie Collector www.collectorz.com
More on page 185

Options for Backing Up and Archiving Video from DVD, Hard-Drive, and Memory Card-Based Camcorders

Output your video from the camcorder to a standalone disc burner. Since your original video is in an MPEG-2 or lower-quality format, you won't lose video quality or editing flexibility by burning to disc. For hard-drive and memory card camcorders, you can make two discs and keep them in different locations, and for DVD-based camcorders, you can separate the original disc from the copy. You might also copy the video to a hard drive and store the discs in a remote location.

Transfer your video to an external storage device or NAS with redundancy such as RAID, or to a Windows Home Server. (Read more about these options in Chapter 1). If you want to store copies remotely to protect your videos from disasters and theft, back up your video files to a small external hard drive or high-capacity optical discs, and keep the backup in a safe deposit box or at a friend's house.

Digital Video Archiving Tips

Avoid FAT32. If you store your video on a hard drive, make sure that it is formatted with a file system other than FAT32, which does not support files larger than 4GB. You can

reformat most drives in a file browser such as Windows Explorer—but be aware that formatting a drive makes its contents impossible to retrieve without data-recovery software or services, and might cause data to be permanently lost.

Use automatic scene indexing. Some video editing programs and specialty software will detect and index scenes in your video when you import it into the software, or in some cases (for example, CassetteDV at paul.glagla.free.fr/cassettedv_en.htm), when you simply run the software on a tape in a camcorder connected to your computer.

Label your cassettes and discs. Do this as soon as you fill them with video and are no longer recording on them. Use a special disc marking pen for writing on optical discs; regular pens can cause damage. Better yet, use software that can automatically generate a disc or cassette case cover from catalog, video chapter, and scene index information.

Create disc menus. If you burn your video to disc, use software that lets you create menus or automatically generates them for you. This allows you to see an overview of the disc's contents when you play it and navigate directly to the scene you want to see.

Use offline cataloging. Some video editing and management programs (see page 133) can add offline media such as optical discs to their catalogs. To catalog offline media, you need to connect it to your computer or insert it in an optical drive while running the software. Then, when the media has been stored again, you will be able to view an index of it in the software's catalog, along with the other video that is stored on your computer. That allows you to find footage more easily, without hunting through tapes and discs.

Online Video Hosting Resources

AtomUploads www.atomfilms.com
Blip.tv blip.tv
Break.com www.break.com
Buzznet www.buzznet.com
Dailymotion www.dailymotion.com
GameVideos www.gamevideos.com
Google Video video.google.com
imeem www.imeem.com
kewego www.kewego.com
Liveleak www.liveleak.com
Metacafe www.metacafe.com
MSN Soapbox video.msn.com
OneWorldTV tv.oneworld.net
Ourmedia www.ourmedia.org
Ovi share.ovi.com
pandora tv www.pandora.tv
Peekvid.com peekvid.com
Revver www.revver.com
sevenload en.sevenload.com
Veoh www.veoh.com
Vimeo www.vimeo.com
Vuze www.vuze.com
Yahoo! Video video.yahoo.com
YouAreTV www.youare.tv
YouTube www.youtube.com

Storing Cassettes and Discs

Make sure to store your digital video cassettes and optical discs in a cool, dark, dry place where they won't be exposed to large temperature fluctuations. Store them in their cases in a vertical orientation. Keep cassettes away from electromagnetic sources such as televisions. If you live in a humid climate, add desiccant packets to the cases or storage box. Don't frequently play tapes and discs intended for archiving; make copies for that purpose. Rewind tapes for storage so that the recorded area isn't exposed, and don't touch the surfaces of discs or magnetic tape.

Sharing Videos Online

Once you have organized your collection of digital videos and integrated your analog movies into it, you can publish videos online to share them with friends, family, and the public. It isn't practical to store or archive high-quality videos online the way photos can be stored in online galleries, because of storage costs and upload times. But the Web still provides a way for you to distribute lower-quality copies.

To share your videos privately, you might want to use an online gallery. Many of the options listed on page 111 can display videos as well as photos. To share your videos publicly, a video hosting site or media-oriented social networking site may be a better option. Many sites also offer the kind of exporting tools described on page 112 so that you can easily distribute your videos to other Web sites or a blog. If you shoot video with a camera phone, you can use a service such as ShoZu (www.shoZu.com) to upload the video directly to the Web, just as you would upload photos.

If your video clip is in a high-quality format, you need to convert it to a more compressed, compact format before you upload it to a Web site. Choose video editing software that provides simple tools for creating Web-ready versions of videos or has built-in uploading tools that convert videos to an appropriate format as part of the upload process. The uploading tools provided by video hosting sites may also be able to convert your videos as they are published to the Web.

Before you share your video on a public site, you may want to add credits, a copyright notice, or contact information. You can do this with titling tools available in most video editing programs. Make it difficult to remove your credits by adding text intermittently throughout the video, in a part of the frame where it's not too distracting. Someone who is determined to use your video without permission will probably be able to remove them, but including them will prevent people from casually swiping your footage.

ANTI-SKATING
45
-10
+10

CHAPTER IV

Music

Introduction:

Unless you're under the age of fifteen, your music collection probably includes more than an iPod full of MP3s. In just a few decades, music has migrated from vinyl records and magnetic tape to optical discs to a profusion of digital file formats.

While each new medium has made it easier to build large music collections and play selections wherever you want to hear them, the innovations have also left a lot of recordings in older formats sitting on the shelf. Fortunately, there are simple ways to escort your older recordings into the digital world, integrating them into a unified music collection that is easy to search, sort, play, and share.

In this chapter, you'll learn how to create digital music files from your records, tapes, and CDs, how to restore order to your digital music collection, and how to play your digital music anywhere you please. You'll also find free and affordable resources that can give you access to your favorite old music, as well as a vast array of newer tunes.

Digitizing Vinyl and Tape

The Goal:

Create digital audio files from your vinyl records and tapes so that you can integrate their contents into your digital music collection, attach album information to them, and easily add them to playlists on portable and home players.

What You Need:

For records: a USB turntable—or a standard turntable with a separate preamp or a stereo receiver with a phono input. For tapes: standard or USB tape deck. Cables and connectors, an audio input jack or USB port on your computer, and audio recording software.

Time Required:

The recorded time of your vinyl records and tapes, plus about five minutes to split, save, and add information to tracks.

If you have a vinyl record collection, there's a good chance it's been sitting quietly on the shelf for some time now as CDs, iPods, and digital music receivers complete their conquest of your home audio landscape. Most of us will probably never go back to the habit of handling and turning each record carefully when we want to listen to music, now that we can hear potentially endless playlists of songs with just one click. If you have a small record collection that you can replace with commercially available digital files for less than $100 or so, it may not be worth the effort to digitize your vinyl. But you should consider it if you have a very large collection, have rare recordings that you can't find digital versions of, or just dig the sound of a needle in a groove. Even if you no longer own a turntable or a stereo receiver with a phono input, there is affordable equipment available that can help you create a digital version of your record collection.

Record Digitizing Setups

The path between your vinyl record and a digital file includes several stops. First, a turntable has to play the record, then the sound has to be equalized through a phono preamp, then it has to be fed into your computer or another digital recording device. There are several setups that can be used to create this path:

✔ **1. All-in-one digitizing turntable.** One of these devices combines a turntable and preamp with a CD drive or a dock for an MP3 player such as an iPod. To digitize a record, you simply play it while recording to the inserted disc or mounted MP3 device. All of the controls are built in, so you don't need a computer at all. This can be very convenient, although recording directly to MP3 isn't the highest quality option. It also requires you to hit a button between songs to create separate tracks. Some of these turntables have automatic track separation features, but they don't work well for all recordings. You may also have to use media player software to add album information later, when you transfer the digital files to a computer.

✔ **2. USB turntable and a computer with a USB port.** USB turntables have built-in preamps and USB ports so that you can simply connect the turntable directly to the USB port on your computer with a USB cable.

✔ **3. Standard turntable, stereo receiver with a phono jack,** and a computer with audio-in jack. In this case, the preamp is built into the receiver. The sound card in the computer provides the input jack. You connect your turntable to the receiver's phono input with a standard RCA cable. Then you connect the receiver's line out to the line-in jack on your computer with an RCA cable. If your computer line in is a minijack, you'll need a jack adapter, which you can pick up at an electronics store.

✔ **4. Standard turntable, standalone phono preamp with USB port** or line-out jack, and USB port or audio line-in on your computer. With this option, you connect the turntable to the preamp with a standard RCA cable, then connect the preamp to the computer's USB port with a USB cable or to its audio line-in jack with an RCA cable (and a minijack adapter if necessary).

When you're choosing equipment, keep in mind that its quality will affect the audio quality of the digital files that result. A turntable with a high-quality needle cartridge, a well-balanced tone arm, and a solid metal platter will give you better sound to start with than a cheap needle and an all-plastic turntable. Likewise, a high-quality preamp and a dedicated sound card will do a better job of preserving that great sound as it makes its way into your computer than a USB connection and onboard sound.

If your main goal is to be able to carry your tunes on an MP3 player or play them on computer speakers, you'll probably be fine with the sound you get from the lower-quality options. But if you're going to add your newly digitized records to a high-quality music collection and play it on your stereo, make sure your hardware is up to snuff. Advanced recording software such as Pure Vinyl comes with detailed information about recording setups for audiophiles.

Unless you're using an all-in-one turntable, you will also need to select audio recording software. USB turntables usually come with it. If you're using a standard turntable, look for a program that is designed especially for digitizing records or provides dedicated tools for that purpose.

The Recording Process in Ten Steps

Once you've decided on a recording setup and hooked everything up, here's what you need to do. The details of this process will vary somewhat, depending on the setup and software you use.

✔ **1. Clean your record.** Use a soft cloth or record brush, or even a record cleaning solution from a specialty store for especially dirty records. Make sure your record and turntable needle are free of dust before you start. (If you're using an all-in-one device, you'll simply follow the manufacturer's recording instructions from this point. When you download the digital tracks to your computer you can go to step 9.)

✔ **2. Check your PC's audio input.** If you're using Windows, go to Control Panel/Sound and make sure the correct audio input is selected.

✔ **3. Open your recording software and check the audio input.** Check the preferences or options and make sure that the correct audio source is selected. If available, select a software playthrough feature so that you can hear the record through your computer's speakers as it's recording. If you don't have this option and you're using a USB turntable, you may need to connect it to headphones or a powered speaker in order to hear it.

✔ **4. Check your recording settings.** Most programs will default to a 44100 Hz sampling rate and 16-bit bit depth. Audiophiles who know their way around audio software and are willing to devote substantial storage space to large files may want to select higher settings. For everyone else, those rates should be just fine. Make sure stereo recording is selected, unless you're recording an old mono record.

✔ **5. Digitize the record.** Hit the Record button in your software, and play the record on your turntable. Your software will display a visual representation of the sound as it records. When the first side of the record is done, turn it over and play the other side.

✔ **6. Remove clicks and pops.** This is an optional step that you can take if your software provides an appropriate tool or filter.

✔ **7. Split the tracks.** This process works differently in different programs, but the general idea is to click on the points between songs to break them up into separate tracks. Some

Audio Recording Resources ▸▸

Recording software makers

Acoustica www.acoustica.com
AIPL www.aipl.com
AlpineSoft www.alpinesoft.co.uk
Audacity audacity.sourceforge.net
Audiotool www.audiotool.net
AVSMedia www.avsmedia.com
Cakewalk www.cakewalk.com
CFB Software www.cfbsoftware.com
Channel D www.channld.com
Coyote Electronics www.coyotes.bc.ca
NCH Software www.nch.com.au
PolderbitS www.polderbits.com
Roxio www.roxio.com
TongSoft www.tongsoft.com
Tracer Technologies www.tracertek.com
Wieser Software www.ripvinyl.com

USB turntable makers

Audio-Technica
www.audio-technica.com
Grace Digital Audio
www.gracedigitalaudio.com
ION Audio www.ion-audio.com
Numark www.numark.com
TEAC America, Inc. www.teac.com

Digitizing Audio Cassettes ▸▸

The reasons for digitizing audio cassettes are much the same as for digitizing vinyl records: If you have a big collection or rare tapes, it might be worth the time required. The digitizing process is also quite similar, but you don't need a preamp to digitize tapes as you do with records. That means you can hook up any standard tape deck directly to your computer, as long as your computer has an audio-in jack. If it doesn't, tape decks with USB output are available. You can connect a standard cassette deck to your computer's audio input jack with an RCA cable (and a minijack adapter if your computer has a minijack).

You will need audio recording software on your computer to capture the cassettes as they play. (Find a program in the Resources list on page 140.) Look for software that has a tool for reducing tape hiss in the captured file. As with vinyl records, make sure the software you select gives you easy ways to separate long recordings into tracks.

The quality of your tape deck and the sound card in your computer will affect the audio quality of the digital files you capture. However, old cassette tapes are not likely to offer very high audio quality to begin with, so think carefully before you run out and buy an expensive deck.

Once you've connected your tape deck to your computer, you can follow the same steps that are described here for digitizing records. Make sure the heads inside the tape hatch of your deck are clean before you start. To clean them, you can use rubbing alcohol on a cotton swab or a special head-cleaning cassette purchased from an electronics store.

software allows you to do this as the recording is made. Otherwise, you will probably need to play the recording back in order to put the breaks in the right places. Some programs let you add silence and fade-outs and -ins, which can be useful for creating natural beginnings and endings for tracks from live recordings.

✔ **8. Export or save the tracks in the format of your choice.** I recommend using an uncompressed or lossless compressed format (read more about audio formats on page 144). You can always create lower-quality copies for use on MP3 players. Look for a program that allows you to export multiple tracks at once, so that you can save the whole album in one efficient process. Save the files to your Music folder so that your media player will find them easily.

✔ **9. Name the tracks.** This is usually part of the process of splitting tracks or exporting or saving your track files. Naming your tracks appropriately is important not only so that you can identify them by file name later, but also because it may help your media player software to fill in other album information later. Read more about naming audio files on pages 149-151.

✔ **10. Add album information.** Some recording programs have this feature built in, but in most cases, you will need to open your newly created audio files in a media player to attach album information from online sources to them. For example, in Windows Media Player, you can right click on the tracks and select Find Album Information to automatically find and attach album information to your tracks. You can also use an ID3 tag editor in your recording or player software to add information manually.

Ripping Discs

The Goal:

Copy your audio CDs to digital files on your computer so that you can integrate them into your digital music collection, attach album information to them, and easily add them to playlists on portable and home players.

What You Need:

A computer with an optical disc (CD) drive, media player software, and an Internet connection.

Time Required:

About five minutes per disc; varies according to recorded audio length and computer, optical drive, and software performance.

If you have CDs and you have a computer, it's unlikely that you don't already know how to rip a disc, in other words, how to copy the tracks on your CD into files on your computer. In fact, it's hard to put a disc in a computer drive these days without some program popping up and offering to rip it. Anyone who has a computer with a current operating system has tools for ripping discs, since this function is included in free media player software that comes with the operating system, such as Apple iTunes and Microsoft Windows Media Player. (If you're interested in exploring your media player software options, check the Resources list on page 185.) However, there are a few things you should know about ripping discs in order to get the quality you expect, and to make sure the music you transfer to your computer is easy to manage and integrate into your existing collection of MP3s and other digital audio files.

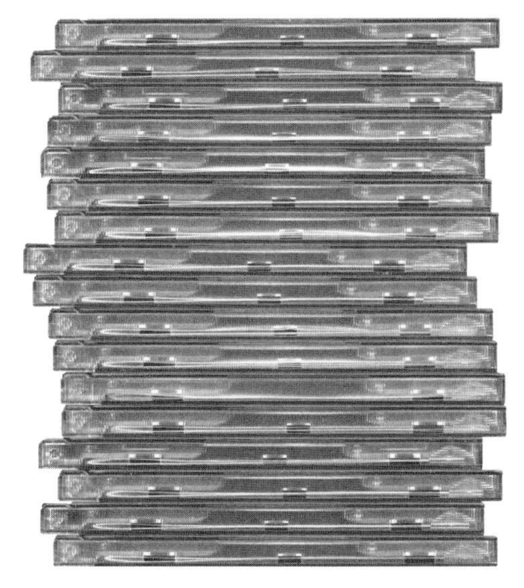

Take these steps when you rip discs to your computer:

✔ **1. Check your drive.** The quality of your CD drive and its firmware can affect the quality of the audio files you rip from discs. Consider replacing an old, generic drive before ripping a large collection of CDs. Check the drive manufacturer's Web site for firmware updates. Also make sure your discs and drive are clean before you start. If you hear glitches in your ripped audio files, try using media player software that offers an error correction option in its disc-ripping settings.

✔ **2. Choose the right format.** Instead of going with whatever format is the default, take a look at the options your software provides and make an educated choice. If you rip a CD to a format such as MP3, you will lose some of its audio quality. If you want to create a high-quality digital archive of the music you copy from discs, you can rip them to a high-quality audio format and then make smaller, lower-quality copies for playing on devices such as MP3 players. Read more about audio formats on page 144.

✔ **3. Choose the right bit rate.** The same points apply as with formats: a low default bit rate sacrifices audio quality for file compactness. Read more about bit rates on page 144.

✔ **4. Get online.** Make sure your computer is connected to the Internet when you rip discs. Most media player software automatically connects to an online source to acquire information about the album you're copying, and then attaches it to your track files. If you're offline, you'll end up with a bunch of unidentifiable files that have generic names such as "Track 1." If you can't get online while you're ripping a disc, or your software doesn't find and attach album information automatically, then look in your software menus for an option to manually activate this function next time you're connected to the Internet.

✔ **5. Turn off copy protection.** Some media players can add copy protection (a.k.a. digital rights management) to ripped audio files. This feature limits the number of devices the tracks can be played on and prevents the files from being copied. To play a copy-protected file on a new computer or device, you have to enter license information, which is then approved by an online license server. Unless you have a reason to limit the use of the audio files you're creating, go into your software's menu of ripping options and make sure that copy protection is switched off.

✔ **6. Check the file destination.** Your software will probably select the right folder to put your newly created audio files in by default, but it's best to check and make sure. Read more about where to store your audio files on page 148.

✔ **7. Rename files.** In most cases you won't have to do this, since the files will be given appropriate names when the software automatically attaches album information to them. However, if you get a CD from an amateur source—from a local musician at a concert, for example—your software might not be able to find relevant information. In this case, you'll have to change the file names from the generic "Track 1," and so on, that your software assigns them. Read more about naming audio files on pages 149-151.

✔ **8. Import or add album information.** When you rip discs, your software will automatically attach album information to the resulting files. However, if it is unable to find relevant information from an online source, or if you simply want to add more information than is automatically available, you can use your software's tag editor to add information. To keep your files organized, do this as soon as you rip discs. Read more about adding album information on page 148.

Understanding Digital Music Formats

The digital music format that most people are familiar with is MP3, but that's hardly the only one available, or even the best one for all purposes. You can identify the format of a digital music file by its extension. For example, MP3 file names end with the extension ".mp3." If you're going to rip discs or digitize vinyl albums and tapes, it's important to know a little about the digital formats that you can select to save your music in. And even if you're just buying digital music online to add to your collection, you may have options other than MP3—or various quality levels of MP3—to choose from.

Higher-quality files can retain more detail and clarity in the range of sounds that they preserve, which means that you'll hear more nuances in the music. Of course, the level of nuance and clarity you hear also depends on the quality of the equipment you're using to play your music. If you're listening to music on a portable player or computer speakers, you probably won't benefit much from having the highest quality files. But if you're using higher-end audio equipment and speakers, and you have a critical ear, you'll notice a difference. And if your goal is to preserve your music collection in a high-quality archive, then choosing the right format is important. As with any type of digital file, increasing quality generally means increasing files size as well, which means that you'll need more storage space for your high-quality files.

Two of the main factors in file quality are compression and bit rates. Compression makes the files smaller—so that you can do things like fit thousands of tracks on your portable player—but it also reduces quality. Formats that use lossy compression (when data is lost during the compression process) are also susceptible to further degradation if they are resaved. If you're making new versions of files in a different format, it's best to start with the highest quality file possible instead of converting a lower-quality version. For example, if you have a WAV file and an MP3 of the same song, and you want an AAC version of it, convert the WAV file to AAC.

Music files can be saved at a range of bit rates, which define the amount of digital data that the file can use to convey sounds in a given period of time. Music file bit rates are described in kilobits per second. As a rule of thumb, don't use a bit rate lower than 128 kbps when you're saving files (and don't select a lower rate when purchasing them). At 128 kbps or higher, most people will be satisfied with the audio quality. At lower rates, even casual listeners may hear a drop in quality. Audiophiles with high-end sound systems may want to choose a bit rate substantially higher—320 kbps or more.

Because the quality of sound depends upon a combination of the file format's characteristics and the selected bit rate, different formats will produce different levels of quality at a given bit rate. Many careful listeners agree that an AAC or Ogg Vorbis file offers higher quality than an MP3 at the same bit rate. But there's a lot of debate. Does a 192 kbps MP3 sound better than a 160 kbps AAC? How about a 160 kbps Ogg Vorbis versus a 160 kbps WMA? The jury is still out. If you want to judge for yourself, rip a CD track to several compressed formats, with different bit rates, and then compare the results. Better yet, have a friend play them randomly for you while you're not looking at which one is being selected, so that you can tell whether you can really distinguish between them. Use a piece of music with a broad range of tones and complex instrumental arrangements, and play your samples back on the best audio equipment you have.

Digital Rights Management

Unfortunately, sound quality and file size aren't the only considerations you need to keep in mind when you're buying digital music. Another big factor is digital rights management, or DRM. This is a means of embedding information in the file to limit its use. Companies that sell music files use DRM to restrict your ability to play the files with software and devices, control the number of times you can copy them, or make the files expire. If a file supports DRM, you may not be able to play it on all devices or players. One of the reasons why MP3 is such a popular music format is that it doesn't support DRM, and can therefore be played just about everywhere, and won't expire. The way that DRM works is by periodically renewing your right to play the file by checking a license stored on an Internet-connected server. This means that if you are able to use DRM-protected files on a portable device, you'll need to sync it with your Internet-connected computer's media player periodically in order to keep the files playable.

So how do you know if the files you're buying have DRM? If you're buying any file type that can support it (see the chart opposite), check the current policy of the company selling the file to see if it's using DRM. If it is, be aware that the limitations the company imposes through DRM could change. For example, the company could announce one day that it has stopped supporting certain DRM-protected files with its devices and software, and stop renewing the licenses that allow you to move them to a new computer or hard drive. That would leave you with nothing on which to play them when your current computer, devices, and software become outdated or need to be replaced. The effect of this would be sort of like someone from one of the big music labels walking into your house one day, taking all of the CDs you bought from that company, and throwing them in the trash. Sound crazy? It can happen, and it already has. So the moral is, find out if the files you're buying are DRM-protected, and understand the risks if you go ahead and make a purchase. You don't have to worry about DRM when you rip, convert, or record files yourself, only when you purchase them.

DRM does have some positive uses. For example, when you subscribe to an online music service that provides access to millions of songs, it may use DRM to let you play music on a portable device. By protecting the files you transfer to your device with DRM, the service can also cause the files you download to expire and become unplayable if you cancel your subscription. That way, you can download as much music as you like to your device while you're a subscriber, but the service isn't simply handing over its vast music collection for a low monthly subscription fee. In other words, DRM can make sense when you're renting music, but not when you're buying.

A Quick Overview of Popular Music File Formats

FORMAT	FILE EXT.	COMPATIBILITY	COMPRESSION/ SIZES	DRM SUPPORT	GOOD FOR
MP3 (MPEG-1 Audio Layer III)	.mp3	Nearly universal	Lossy compression, small file size	No	Portable players, any compatible playback device unless highest audio quality is desired
AAC (Advanced Audio Coding)	.m4a, (.m4p for DRM)	Apple products, some third-party players and devices	Lossy compression, small file size	Yes	Portable players, any compatible playback device unless highest audio quality is desired
WMA (Windows Media Audio)	.wma	Microsoft products, some third-party players and devices	Lossy compression, small file size (lossless option also available)	Yes	Portable players, any compatible playback device unless highest audio quality is desired
Real Audio	.ra	Real Networks products, limited number of third-party products	Lossy compression, small file size	Yes	Portable players, any compatible playback device unless highest audio quality is desired
Ogg Vorbis	.ogg	Limited	Lossy compression, small file size	No	Portable players, any compatible playback device unless highest audio quality is desired
WAV (Wave Form Audio)	.wav	Broad	Uncompressed, large file size	No	Archiving, ripping discs, digitizing records and tapes, high-quality playback
AIFF (Audio Interchange File Format)	.aiff	Apple products, some third-party players and devices	Uncompressed, large file size	No	Archiving, ripping discs, digitizing records and tapes, high-quality playback
FLAC (Free Lossless Audio Codec)	.flac	Limited	Lossless compression, moderate file size	No	Archiving, ripping discs, digitizing records and tapes, and high-quality playback when a smaller file size than WAV or AIFF is desired

Organizing Your Digital Music

The Goal:

Organize your digital music files into a folder system, identify and rename mystery files, and add album information to your tracks.

What You Need:

Your digital music collection on your computer and any attached or networked storage devices, and software with music organizing tools.

Time Required:

Depends on the software tools you select and the number of poorly named and organized tracks you have.

Most people use their favorite media player software to organize their digital music collections, and that's the approach that makes the most sense. The software is free or inexpensive, and it gives you many more tools for managing your music than something like a file browser.

The key to using media player software to organize your music files is metadata. The metadata are all of the information about your music that is attached to each track's file in areas that are usually referred to as tags. Tags can include basics like the artist and album names, as well as elements that are nice to have but not essential, such as song lyrics. Without metadata, your music tracks become anonymous files with names like "Track 1" that are impossible to sort, search, and identify without playing.

When you purchase digital files such as MP3s online, you don't have to worry about metadata—they will already be attached. But when you rip discs, digitize records and tapes, or acquire music files from amateur recording sources, you have to make sure that the resulting files are named properly and have some basic information attached. If you're connected to the Internet when you rip discs, your media player software should be able to automatically find the relevant metadata online and attach them to your files. In the recording software that you use to digitize analog music, you will be given the opportunity to name files and add information when you save the files. But we all know that somehow, somewhere along the line, we end up acquiring unidentifiable tracks—sometimes a lot of them.

Obviously, one way of organizing unidentifiable files would be to listen to each one, rename it manually, and type in the information that should be attached to it by using your media player's tag editor. If you have that kind of time on your hands, you probably don't need to be reading this book. A much more efficient approach is to use automatic tagging and file naming tools. These work in conjunction with a built-in or online database of music metadata, and can be found in advanced media players and specialized software programs. The tagging tools look at the contents of each file you select, find a file that is identical to it in a database, and then copy the metadata attached to the file in the database to your file. Then the file naming tools can automatically rename your file by using information in the metadata and a naming scheme that you designate. For example, if you set files to be named with "Track Number Track Name_Album Name_Artist," the software will search the metadata and rename your file "01 Fidelity_Begin to Hope_Regina Spektor."

While some tagging software has a built-in database to cover a selection of the most popular music, the online databases that automatic tagging tools use have much more information, so it's important to be online when you use the tools. The databases that autotaggers search include Gracenote (sometimes still referred to by its former name, CDDB), MusicBrainz ▼, freedb, and AMG, and some tools also search sources such as Amazon and Apple to acquire album cover art and other elements. Each software program's system for automatic tagging and renaming works a little differently. Popular (and free) media players such as iTunes and Windows Media Player tend keep things simple with one-click tools that offer to "Get CD Track Names" or "Find Album Information," for selected tracks, then let you choose the album to which your tracks belong from a list of possible database matches. Others provide more complicated interfaces with more powerful tools. Yet another option is to enhance your free media player with a powerful automatic tagging plug-in such as TuneUp, which can be added to iTunes.

Johnny Cash
1932-02-26 - 2003-09-12, Type: Person, Default data quality
Cash, Johnny
Info: [Link to this | Details | Aliases | Tags | Releases | Appears on | Similar artists | Search google]
Edit: [Log in to edit this]

America has Amazon ASIN B00005U2LX [info]
[View relationships]

Yellow highlights indicate pending changes.

America
Release: [Link to this | Details | Tags | Show artists | Hide relationships | Search google]
Edit: [Log in to edit this]
Tags: (none) - Tag this release
Amazon: info | buy

#	Track	Length
1	Opening Dialogue PUID: TRM	0:24
2	Paul Revere PUID: TRM	2:18
3	Begin West Movement PUID: TRM	0:27
4	The Road to Kaintuck PUID: TRM	1:22
5	To the Shining Mountains PUID: TRM	0:49
6	The Battle of New Orleans PUID: TRM	2:21
7	Southwestward PUID: TRM	0:38
8	Remember the Alamo PUID: TRM	2:27
9	Opening the West PUID: TRM	0:56
10	Lorena PUID: TRM	1:50
11	The Gettysburg Address PUID: TRM	2:38
12	The West PUID: TRM	0:58
13	Big Foot PUID: TRM	3:02
14	Like a Young Colt PUID: TRM	0:38
15	Mister Garfield PUID: TRM	3:54
16	A Proud Land PUID: TRM	0:25
17	The Big Battle PUID: TRM	3:05
18	On Wheels and Wings PUID: TRM	0:31
19	Come Take a Trip in My Airship PUID: TRM	1:06
20	Reaching for the Stars PUID: TRM	0:42
21	These Are My People PUID: TRM	2:39

21 1 63 PUID: 301 TRM [Default data quality] | [English, Latin] | [Album, Official]
Disc ID: LD_L17kFcmZXe39D7HhpmhqiaEY- 33:10

Date	Country	Label	Catalog #	Barcode	Format
1972	United States	-	-	-	-

Media Player and Management Software Resources ▸▸

There are dozens of programs that let you import or edit music metadata, find duplicate files, and play music. These are some of the most popular and useful.

Apple iTunes www.apple.com
Duplicate Music Files Finder www.lcibrossolutions.com
Media Catalog Studio www.maniactools.com
Media Jukebox www.mediajukebox.com
MediaMonkey www.mediamonkey.com
Microsoft Windows Media Player www.microsoft.com
MP3tag www.mp3tag.de/en
Music Label www.codeaero.com
Music Library www.wensoftware.com/MusicLibrary
MusicBrainz Picard musicbrainz.org
Teen Spirit teenspirit.artificialspirit.com
The GodFather (available from www.download.com)
TuneUp www.tuneupmedia.com
Winamp www.winamp.com

If you have a lot of unidentifiable music tracks, it's worth the time to learn how to use one of the more powerful options. MusicBrainz Picard is a good example. It will search your system for music tracks that are missing metadata and are poorly named, scan the files to analyze them, then look them up in the MusicBrainz and freedb databases to download information about the entire album in which each track belongs. To tag your tracks, you simply drag them onto the matching track information that has been downloaded. Then Picard fills out the missing metadata, renames the files, and sorts them into new album folders for you.

Other advanced features that are very useful include de-duping tools and moved-file locators. De-duping tools automatically detect files that may be duplicates, then let you rename or delete them. Tools that locate moved files will automatically search for the music that is listed in your software's catalog but cannot be found in the folder where it was first cataloged. The software will then record the files' new locations in its catalog so that the tracks can be played.

It's unlikely that you'll be able to clean up all of your music files with automatic tagging and file naming tools. You'll probably have to handle a few manually. To do this, use your software's tag editor. Music metadata are usually attached to files in the ID3 or ID3 version 2 format (ID3 version 2 accommodates more information than the older ID3 standard). Select the file you want to fix up, then open the tag or ID3 editor and fill in some basic information, such as the track title, artist and album names, and track number. You can fill in additional information if you want to, but having the basics will allow you to search, sort, and identify the file. After you've filled in the metadata, use an automatic renaming tool to rename the file according to the information in the metadata, or, if your software doesn't have an automatic renaming tool, use a file browser to rename the file.

Once you have tagged and renamed your files, you can use automatic tools for sorting them into folders. If you've used a powerful auto tagging tool such as Picard, it may have already performed this task as part of the tagging and renaming process. If not, look for an automatic organization function that creates new subfolders in your Music folder (or any other umbrella folder that you designate) and sorts your files into them. You should be able to specify the folder structure that you want the automatic organizer to use. A simple folder structure such as Music/Artist/Album works well. Not all music player software offers this kind of automatic folder organization tool, but if you have a disorganized music collection, it's worth spending a little money on a program that does (MediaMonkey is a good example). The alternative is creating artist and album folders yourself and sorting your files into them manually—a tedious and time-consuming job if you have thousands of tracks.

If at the end of the organizing process there are a lot of empty folders on your system, you can download and run a program such as Remove Empty Directories (www.jonasjohn.de/lab/red.htm) to detect and delete them. Don't delete random empty folders if you don't know what they're for, however. Some programs create empty folders that need to be available for the program to make use of in the future.

Naming Music Files

When you're renaming music files, it's best to keep things simple. The track number and name are generally the most important pieces of information to include in the file name. You can also include the artist and album name if you want, although if you add this information to the file metadata, it will appear in your media player without you having to create extremely long file names.

Creating Playlists

Playlists are another organizational tool, and you can spend as much or as little time as you want creating them. Playlists are just what they sound like—lists of tracks that will play continuously in the order you choose. Almost all media player software lets you create playlists by dragging and dropping music files onto a list and then rearranging their order as you like. Many programs can also automatically generate playlists according to criteria that you select. Creating playlists is a good idea not only because it makes your music easy to play at home, but also because you can load your playlists onto portable players and share them online.

Using Subscription and CD Ripping Services

Having your music collection available in digital form can make it much more accessible and convenient to listen to. Once you're dealing with digital files, you can copy them to portable players and cell phones with built-in music players, stream them to digital receivers in your home, and create playlists so that you don't have to keep pushing buttons or changing discs to keep the music playing.

One way to bring all of your music together in a unified digital collection is to rip your CDs, digitize your vinyl records and cassette tapes, and integrate the files you create into the collection of MP3s and other digital music files that you already have. That approach requires a substantial amount of time (or money, if you decide to have your music ripped and digitized by a service), as well as a potentially large amount of hard drive storage space. If you don't have a turntable or a tape deck anymore, this approach might also require you to purchase some audio equipment.

But there's another way to accomplish the goal of gaining access to your music collection in digital form: Use a subscription service. Not only will the service let you play most if not all of the music that can be found in your own collection, but it will give you access to millions of additional tracks.

Napster (www.napster.com) and Rhapsody (www.rhapsody.com) were the two main streaming music subscription options at the time this book was written, although by the time you read it, additional options may exist. The way these services work is that you pay a monthly or annual fee and receive access to a massive collection of digital music, which you can search and listen to on a computer or play on compatible receivers and portable devices.

Internet Radio ▸▸

Another way to gain access to large collections of music is through Internet radio, and it's usually free. Unlike regular radio, Internet channels can hand the DJ headphones over to you, giving you control over the specific artists or albums that are played. You can't be as precise in selecting tunes and albums that you want to hear as you can with a subscription service, but if you want to listen to music by a single artist or group of artists, you can find a way to set that up. Radio services like Pandora (www.pandora.com), Soundpedia (soundpedia.com), and Slacker (www.slacker.com) give you highly customizable ways to create "stations." Social networks like imeem (www.imeem.com), Deezer (www.deezer.com), MOG (www.mog.com), and Last.fm (www.last.fm) serve a similar purpose by letting you search for and listen to music that has been shared by other people. To peruse an extensive list of Internet radio options, go to Live365 (www.live365.com) or Radio-Locator (www.radio-locator.com).

In most cases, you download an application to your computer when you subscribe to the service, but sometimes you can just use a Web browser to play music from the service's collection. The desktop software that subscription services provide works much like typical media player software, allowing you to add tracks to your own library and playlists. The difference is that the tracks aren't actually stored on your computer; when you click Play, the music is streamed over the Internet to your computer. Signing up and installing any necessary software takes just a few minutes, and because you don't actually download the music, using the service requires a fairly small amount of hard drive space.

You may wonder whether this type of service would really have all of the same music that you own in your personal collection. Unless you're an avid collector of obscure recordings, there's a good chance that they'll have a high percentage of the recordings you own. And you might discover that your "rare" recordings aren't quite as hard to find as you thought. Even if there are a few items that you can't find through a subscription service, digitizing just the portion of your collection that isn't available might be a much less onerous task than handling the whole thing.

There are, of course, some downsides and requirements. First of all, you have to have a fast, reliable broadband Internet connection for the service to work. And you do have to pay a fee—these services aren't free. If you cancel, you don't keep any of the music you could listen to when you were a subscriber (unless you paid extra to download particular tracks or albums). Also, streaming music from a service doesn't offer the absolute best audio quality. It sounds very good, and will satisfy the vast majority of listeners, but it might not be the right choice for serious audiophiles with high-end digital stereo equipment.

Ten Factors to Consider When Choosing a Music Subscription Service

✔ **1. Subscription and download costs.** Annual subscription fees generally start around $150. Some services offer limited free versions and more than one level of subscription service. In addition to listening to streaming music from the online services, you can purchase and download tracks to store at home, which you can keep if you cancel your service. Download costs vary, so if you think you will purchase tracks or albums, compare download pricing before when choosing a service.

✔ **2. Catalog.** Rhapsody and Napster both offer catalogs of millions of tracks, and are continually adding to their collections. Try a free trial period and search for your favorite tunes as well as some of the more obscure music that you want to listen to, to find out which service's catalog best accommodates your taste.

✔ **3. Portable device compatibility.** Music services offer subscription plans that allow you to download music to portable devices such as MP3 players and smartphones. The way this works is that you connect your device to your computer periodically (usually once a month) to renew the rights you have to listen to it. If you don't sync your device with the service this way, or if your subscription expires, you won't be able to listen to the tracks anymore. Check each service to see if it is compatible with a portable device that you own or might buy.

✔ **4. Digital receiver compatibility.** Music services are also compatible with some digital audio receivers, so that you can stream music to a stereo, TV, or other player on your home network. Check the services to see if they support any receivers you might own.

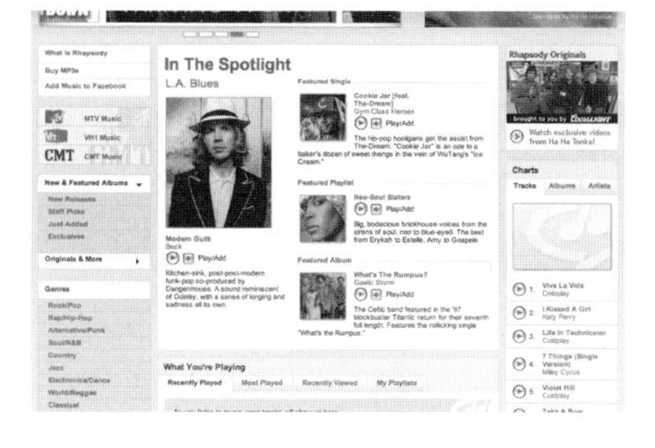

✔ **5. Music management tools.** The software that you download lets you create playlists of the music you listen to, provides flexible tools for searching for and sorting music, and can generate playlists automatically. It should also offer tools for managing your local music collection so that you can integrate your rented tracks with the ones you own to create combined playlists. Look for a service with software you find easy to use and flexible enough to organize your music the way you want to.

✔ **6. Editorial features.** One of the advantages of using a subscription service is that you get access to information about the music and musicians you like. Look for features like editorial reviews and concert information.

✔ **7. Community features.** Look for features that let you share your playlists and music with friends or others with similar taste, and tools for publishing your playlists online.

✔ **8. Download file quality.** If you think you'll purchase music for download from the service you use, check the file formats and bit rates available. Also find out whether files for download are protected by digital rights management.

✔ **9. Internet radio and custom channels.** Some services provide access to Internet radio and the ability to create custom "channels" to play from the service's desktop software.

✔ **10. Recommendations.** The software you download from a subscription service typically provides ways to get listening recommendations based on the music you've already listened to, or to find out what other people with similar tastes are listening to. This can be an excellent way to learn about new music that you wouldn't otherwise come across.

Getting Your CDs Ripped by the Pros

It's neither difficult nor expensive to rip your own CDs. However, if you'd rather spend money instead of time to accomplish the task, you can send your discs to a professional CD ripping service. These services charge about $1 per disc.

All you have to do to have your CDs ripped professionally is select a service (see the Resources list on page 186), place an order online or by phone, pack up your discs, and ship them. The service will send your ripped files back to you on a hard drive or data DVDs, along with all of your CDs. The ripping process usually takes about a week, and some companies offer express service. Here are some features to look for:

✔ **1. Disc cleaning.** This should be included in the cost of ripping a CD.

✔ **2. Repair and resurfacing of scratched discs.** Many services also include resurfacing of discs with physical damage that affects playback.

✔ **3. High-quality audio formats.** You should be able to select from a range of music file formats, rather than being limited to MP3s.

✔ **4. Selectable bit rates.** If you choose to have your CDs ripped to a format that can be saved at numerous bit rates, you should be able to select the bit rate.

✔ **5. High-quality discs and hard drives.** Ask about the brand and type of DVDs on which your files will be saved. If you purchase a hard drive from the service instead of sending an empty drive to have your files stored on, make sure you're getting it at a competitive price.

✔ **6. Included metadata input and verification.** Make sure that the service you select will tag your ripped files with information about the music and will verify that its automatic tagging system has provided accurate information for all files. The service should add any missing information manually.

✔ **7. Album art.** If any album art is available, you should receive it with your ripped music files so that the album art can be displayed by your media player software or device when you browse and play your music.

✔ **8. Free shipping and packaging.** Most services include shipping and CD packaging in the ripping cost. When you place an order, the service sends you the packaging. The catch is that the packaging usually includes a spindle to hold your CDs, which means that you have to take them out of their cases before shipping them. Make sure the service provides an adequate amount of shipping insurance per disc.

✔ **9. Portable player loading.** Some services offer to load your ripped files onto a portable MP3 player that you can send with your CDs or purchase from the company.

✔ **10. Indexing.** Look for a service that creates an index of your ripped files so that you can locate them easily in your set of data DVDs.

Streaming Music to Your Stereo

Once you've brought all of your music together in a digital collection stored on your computer or an attached hard drive or NAS, you can stream it to any room in your house with the right equipment. Here are a few options to consider as you plan your setup.

Equipment and Setup Options

Wireless RF music transmitter/receiver. This is a simple device that transmits music by radio frequency (RF) from your computer to a stereo, TV, or pair of powered speakers. It comes in two parts: an RF transmitter and an RF receiver. To use it, you plug the RF transmitter into a USB port on your computer, then use an RCA cable to plug the RF receiver into an audio-in jack on your stereo receiver, TV, or speakers. Your computer has to be turned on in order for the music to be available, but most systems allow you to control music playback with a remote control in the room where you have the receiver. You don't have to have a home network set up to use an RF system, since it provides its own transmission components. Basic wireless RF music transmitters cost well under $100. This isn't a solution for audiophiles, but it can offer reasonably good audio quality, and is simple and cheap.

Network music player. This is a higher-end version of the simple music transmitter/receiver. It generally incorporates higher-quality audio components and is made for use with a home wireless or wired network. The cost for network music systems starts around $200 and can run into the thousands of dollars for high-end components in multiple rooms, with options to play different music simultaneously in different rooms. Companies such as Sonos, Roku, and Slim Devices make network music player systems, as do some traditional audio component makers.

To set up a system, you connect its transmitter component to your home network router, then place the receiver wherever you want to hear music. If you're using a wireless system, you don't need to make any physical connection between the transmitter and the receiver, although some receivers can also connect to a wired network via an Ethernet cable.

If you want to place the receiver in a different room than the transmitter and still use Ethernet instead of a wireless connection, you can do so by buying a power-line Ethernet kit. You connect one component of the kit to your network router

Digital Music Receiver, Transmitter, and Player Makers ▸▸

Apple www.apple.com
Buffalo Technology www.buffalotech.com
Choice Select www.choiceselectonline.com
Denon www.denon.com
D-Link www.dlink.com
Freecom www.freecom.com
Hauppauge www.hauppauge.com
Linksys www.linksys.com
Linn www.linn.co.uk
Logitech www.logitech.com
Netgear www.netgear.com
Onkyo www.onkyo.com
Panasonic www.panasonic.com
Philips www.philips.com
Pioneer www.pioneerelectronics.com
Roku www.roku.com
Slim Devices www.slimdevices.com
Sonos www.sonos.com
Sony www.sony.com
X10 www.x10.com

and plug it into an electrical outlet, then connect a second component to the Ethernet port on the receiver and plug it into an outlet. The Ethernet connection is carried along your electrical wiring. This can be useful if you're streaming large, uncompressed audio files to your receiver.

Some receiver components have built-in speakers, although many must be connected to separate ones. Some network music players can also receive music from streaming music subscription services, Internet radio, or satellite radio, even when your computer is turned off.

Audio or A/V receiver with streaming. There's some overlap between the network systems described above and a network audio or audiovisual (A/V) receiver with the ability to receive streaming audio. The important distinction is that a network audio receiver is a full-fledged stereo receiver, with inputs for a wide range of music sources, including CD and tape players, turntables, portable MP3 players, and (sometimes) satellite radio. Audio receivers handle audio sources only, while audiovisual (A/V) receivers can also output video sources to a television. Audio and A/V receivers that support audio streaming have Ethernet connections that allow you to connect them to a home network router and stream music from a computer or NAS. These types of receivers are usually made by the traditional stereo system component makers, with prices that start around $300.

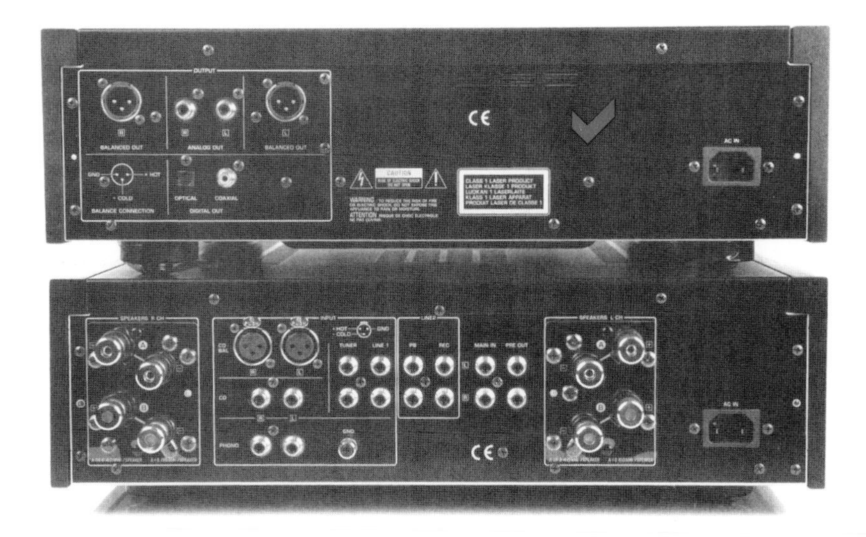

Gaming console or digital video recorder. While it wouldn't make sense to buy a gaming console or a DVR primarily for the purpose of streaming music, if you already have one and your TV has an audio system that you'd like to listen to music on, it's worth checking to see if your console or DVR supports audio streaming on a home network. An Xbox, Playstation, or Wii gaming console, for example, can be connected to a home network via Ethernet, and can stream music from compatible media player software on a computer. The main purpose of a DVR is to record TV shows and movies, but some DVRs can also receive streaming audio from a home network through an Ethernet connection. There are even models, such as TiVo, that can stream music from online subscription services when your computer is turned off. Mac owners who have an Apple TV box attached to their television can set the iTunes media player on their computer to stream music wirelessly to it.

Taking Music With You

There's no mystery to the process of taking your music with you these days: Get an MP3 player, connect it to your computer via USB, and click on the option in your favorite media player software to sync your portable device. If you're reading this book, there's a good chance you already make regular use of an iPod or another portable media player, or listen to music on a player built into your cell phone. However, there are some additional options for taking your music collection with you that you may not be aware of, and there are also a few things to keep in mind to make your musical roving more enjoyable.

Mobile Options

Mobile Internet radio. Some portable devices such as compatible cell phones can receive Internet radio (read more about that on page 152). Radio sources such as Slacker (www.slacker.com) even sell dedicated portable Internet radio devices. FlyTunes (www.flytunes.fm) provides a downloadable Internet radio player that you can install on a mobile device such as an iPod or iPhone. You generally need to have a data service plan or a WiFi connection on your device to receive Internet radio, although Slacker's portable players can store Internet radio downloaded via USB for playback later.

Mobile music subscription service. Some streaming music subscription services (read more about them on page 152) let you download tracks to compatible portable devices. Check the subscription service's Web site to find out which devices are compatible. You don't have to buy the tracks; the portable capability is included in the monthly subscription fee. To keep the music files you download active and able to play, you generally have to connect your device to your computer via USB and sync it with the service once a month.

Online streaming. Some desktop media player makers provide online services that you can use to play your home music collection from a remote computer. The home computer or NAS where you have your collection stored must be turned on and connected to the Internet for this to work. There are also online storage sites that are dedicated to storing music files so that you can play them back on the Web. Winamp's remote service (https://winamp.orb.com) and MP3tunes (www.mp3tunes.com) are examples of these services. Some of the online backup and sync services listed in Chapter 1 also provide built-in players in their Web interfaces so that you can play tunes that are stored on your home computer or NAS or are backed up online.

Mobile streaming. Some of the services and sites that offer online streaming can also stream music to a compatible mobile device such as a cell phone. To use mobile streaming, your mobile device must have a data service plan or a WiFi connection.

Tips on Taking Music With You

Organize your collection before you sync. Put your musical house in order before you sync all your music with your portable player. Sifting through disorganized folders and playlists on a portable device isn't a lot of fun.

Make MP3 copies of high-quality formats. If you have music files in formats such as WAV or AIFF in your collection, make copies of them in the MP3 format or another compressed format that results in smaller files sizes. You are unlikely to hear any difference in sound quality on your portable player, and you will be able to load many more files onto your device much more quickly.

Sync your playlists. ▶ Don't just sync your library of tracks to your portable player. Copy your playlists over from your desktop media player software too, so that your music will play in the order you want to hear it.

Update your firmware. Make sure that your portable player is using the most recent version of its firmware. Check the support section on the manufacturer's Web site for updates. If one is available, installing it should be a simple matter of downloading it to your computer, then transferring it to your device via USB.

Check your file formats. If you have trouble syncing your digital music collection with your portable player, the problem could be that the player does not support the file format that your music has been saved in. Check your player's specs to see what formats it supports, then convert the unsupported files to a supported format with desktop media player software that offers a conversion function. If you have music files that are protected by DRM (read more about that on page 145), you may have difficulty converting or copying the files to play them on a non-compatible device.

PASSPORT
E PLURIBUS
UNUM
United States
of America

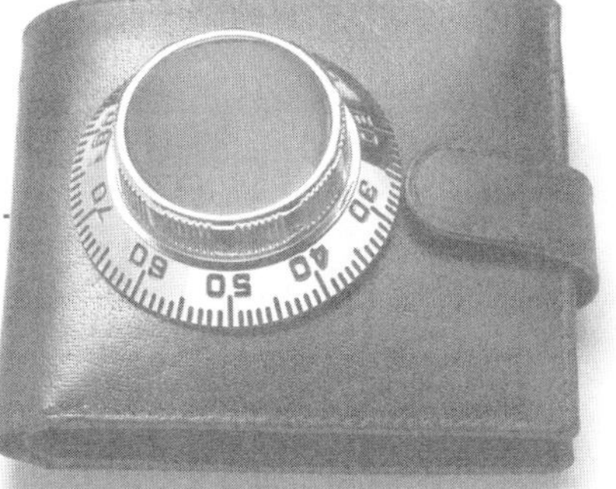

CHAPTER V

Valuables

Introduction:

Drive to the store and buy some groceries with a credit card, and you've just relied upon at *least* five different documents. (Give up? They are: your credit card, your driver's license, your vehicle registration, your title to your car, and your car insurance card.) If you happen to lose your wallet, or return to find that your home has been burglarized or struck by a disaster, you're in for some quality time with bureaucratic paperwork.

While nothing makes the loss of such items pleasant, having copies of your important documents and records of your valuables can greatly reduce the pain. You can use digital technology to secure your documents and create proof of ownership of your valuables. And, on a brighter note, digital tools also give you ways to catalog personal collections for pleasure—or for profit, if you're selling items online.

In this chapter, you'll find tips on how to use digital imaging and even audio recordings to catalog your valuables and documents. And you'll find tools for organizing your digital records, keeping them safe, and making them accessible wherever you need them.

Scanning and Sorting Important Documents

Creating digital copies of your important documents is a good idea for several reasons: First, you can make sure they're preserved by storing the digital files online or on a disc or USB flash drive that you can keep in a safe deposit box or another secure location. Or, you might want to keep the original papers in your safe deposit and store the digital files on your computer to keep them accessible. Second, having your documents in digital form gives you ways to access them anywhere you might need them, either by carrying them with you or by storing them in an online location that you can access via the Web or your cell phone. When your documents are in digital form, it's quick and easy to e-mail them to anyone who needs to see them. And you can keep your digital documents on hand without making them accessible to others, by using password protection and other security measures.

The easiest way to scan your important documents is by using a scanner with an autofeeder that can handle multiple pages. All-in-one printers typically offer this feature, and usually come with document scanning software and features that can make the process of digitizing your documents quick and efficient. You can buy a good all-in-one that includes printing, scanning, copying, and faxing functions for less than $100.

Use the Personal Document Checklist on page 164 to help gather up all the documents you need to scan. If you need to carry an item on the list on an upcoming trip, or if you're concerned about property loss from a possibly impending disaster (if it's hurricane or wild fire season, for example), check the Urgent box to help set your priorities. As you scan each item, check the Scanned box.

Ten Document Scanning and Organizing Tips and Options

✔ **1. Keep your important digital documents in one place on your computer.** When you scan paper documents, save them all to your Documents or My Documents folder. Create a subfolder called something like "Important Documents" if you want, but keep things simple and don't leave documents scattered around your folder system.

✔ **2. Use clear file names.** When you save your scanned documents, make sure you give the files simple names that will be clear to you and anyone else who might use the documents in the future.

✔ **3. Scan both sides.** Don't forget to check the back of documents and cards when you scan them. Some all-in-one printers offer automatic two-sided document scanning, which speeds up the process of scanning a large number of two-sided documents. Save the front and back sides of a document in a single multipage file if your software gives you that option.

✔ **4. Save documents as PDFs.** Look for a scanner that can scan to PDF with the touch of a button or through a simple software interface. This allows you to save your scanned documents in the widely supported PDF format, and to include multiple pages in a single PDF file. You can protect a PDF by making it read-only and setting a password for opening the file when you save it. If your scanner doesn't have a direct-to-PDF function, you can still save copies of image files such as JPEGs in the PDF format if your image editing software offers that option. Adobe's Acrobat.com site lets you create and store a limited number of PDFs online.

✔ **5. Save documents as JPEGs for storage** in online photo galleries. If you have an online photo gallery, you can use it to store important documents if they're saved in an image file format. That gives you both remote access and a secure backup. Just make sure that the site you're using provides good security, which means a private, password-protected gallery.

✔ **6. Use your digital camera.** If you already have a digital camera and don't want to spend more money on a scanner, you can photograph your documents. Use a text mode if your camera has one, and turn off the flash to avoid creating hot spots on the document. Use bright room or natural lighting instead. Fill the frame with your document, and set the camera's white balance to match your light source so that white paper looks white. Make sure your documents are in focus from edge to edge when you photograph them. Use a macro focus mode if you can't get close enough otherwise, and use manual focus if the autofocus won't lock on the document. If you have a lot of documents, consider putting the camera on a tripod and setting up a document easel in front of it so that you can photograph a series of documents quickly.

✔ **7. Use your camera phone.** You can even use your camera phone to scan documents. If it's a smart phone, download mobile scanning software from Qipit (www.qipit.com) or ScanR (www.scanr.com). This type of mobile tool can improve the quality of text images, let you save your scans in PDF format and online, and send the documents you scan to an e-mail account or fax machine.

✔ **8. Organize documents with OneNote.** Microsoft's OneNote software provides a useful way to file important documents, photos, notes, and various other digital items. It works like a digital file folder, allowing you to clip all kinds of different items together and sort them into folders in an intuitive way.

✔ **9. Organize documents with Yep.** If you use a Mac and save your document files as PDFs, this little program can help you manage them in a way that's similar to the way that iTunes handles music. Download it from Ironic Software (www.ironicsoftware.com/yep).

✔ **10. Use a card or receipt scanner.** This scanner can handle a large number of cards or receipts quickly, and its software can automatically add information to a spreadsheet or to a program you use to store contact information. Some scanners also come with online backup service. Card and receipt scanner makers include CardScan (www.cardscan.com), I.R.I.S. (www.irislink.com), and Neat Receipts (www.neatreceipts.com).

▶▶ Personal Document Checklist

DOCUMENT TYPE	URGENT	SCANNED
Appraisals		
Bank account information		
Birth certificates		
Business cards		
Credit card information		
Death certificates		
Diplomas		
Disability insurance documents		
Divorce documents		
Driver's license		
Family recipes		
Frequent flier account records		
Health care proxies		
Home owner's insurance documents		
Immigration documents		
Investment accounts		
Letters		
Letters of recommendation		
Life insurance documents		
Living will		
Loan documents		
Marriage license		
Medical records		
Medical insurance documents		

DOCUMENT TYPE	URGENT	SCANNED
Medicare card		
Mortgage documents		
Passports		
Power of attorney forms		
Prescription information		
Property deeds		
Receipts		
Rental lease		
Resume/CV		
Retirement and 401K account information		
Social Security card		
Social Security documents		
Tax records		
Vehicle insurance documents		
Vehicle titles		
Veterinary records		
Warranties		
Will		

Storing and Accessing Files Securely

Once you've digitized your important documents and cataloged your valuables, it's important to store the digital files securely. You can also take advantage of their digital format to carry them with you when you travel, keep copies in a safe deposit box, or store them securely online—allowing you to access them from any location with an Internet-connected computer or a mobile device.

Ten Tips on Storing and Accessing Important Files

✔ **1. Set your important documents to be read-only.** Use the options in your word processing or other document software to set your files to read-only status before you save them, so that no one can make changes to your important documents.

✔ **2. Password-protect your files.** You can do this in some document creation software, such as Adobe's Acrobat programs for creating PDFs. Make sure you use a password that you'll be able to remember if you have to open the document years from now.

✔ **3. Encrypt your files.** You can encrypt your files with some document creation software or with special file-security software. Encryption encodes the files so that no one else can read them. Opening encrypted files requires a password or a special digital key. Look for encryption options in your software's security or file properties menus. If you're storing your files online or at a remote location that isn't secure, or if you want to carry them on a memory card or USB flash drive that doesn't offer its own security system, encrypting individual files is a good idea.

✔ **4. Store your files online.** If you keep your important files in online storage, they'll be protected from local storage failures and accessible from any Internet-connected computer. Make sure the service you use provides a password-protected storage area and encrypts your files when transferring them. Read more about online storage options on page 40.

✔ **5. Keep digital photos of your valuables in your online gallery.** If you have an online image gallery, you can upload photos and even videos of your valuables to an album there for safekeeping. Use a private, password-protected album for this. You can also store JPEGs of the important documents that you've scanned, giving yourself a way to access them from any location with an Internet-connected computer. However, you should only store documents this way if you're absolutely confident in the security and privacy provisions of the gallery site you're using. Read more about online image galleries on page 108.

✔ **6. Keep your files in online storage that is mobile-accessible.** If you use an online storage service such as SugarSync that gives you mobile access to stored

Secure USB Flash Drives ▸▸
IronKey www.ironkey.com
Kanguru www.kanguru.com
Kingston Technology www.kingston.com
Lexar www.lexar.com
SanDisk www.sandisk.com

documents ▶, you can view them on your smartphone. You'll need to have a mobile data service plan for this to work, and you should be careful about setting your phone to enter your password automatically, in case you lose the device.

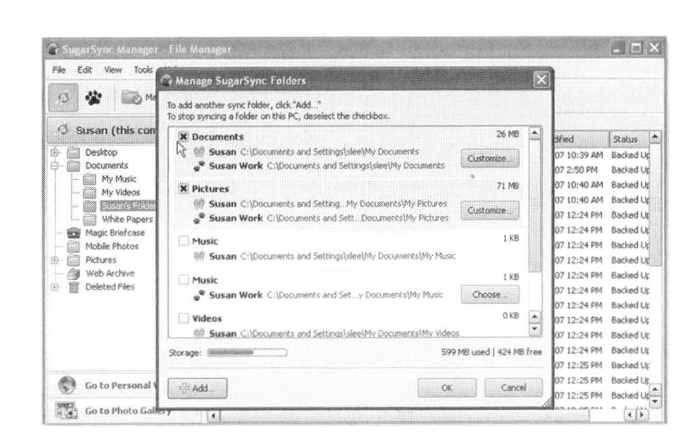

✔ **7. Store your files on your phone.** If you have a reasonably large cell phone memory or a phone that uses mobile memory cards, you can store important document files and images of valuables directly on your phone so that you always have them with you. However, you should only do this if you have encrypted or password-protected the files so that they can't be opened by others if your phone is lost or stolen. You can also install mobile security software on some phones to protect their contents. Security software companies such as Trend Micro (us.trendmicro.com) make programs that you can use to protect files on your phone.

✔ **8. Use FIPS-certified or biometric flash drive.** USB flash drives with Federal Information Processing Standard (FIPS) 140-2 certification offer a level of file encryption that meets the standard used by U.S. and Canadian government agencies for securing documents. A FIPS-certified drive can be especially useful for carrying important documents while you're traveling, since no one will be able to view its contents if it's lost or stolen. If you're forgetful about passwords, though, keep in mind that you won't be able to break into it either if you forget yours. After a certain number of tries with incorrect passwords, some FIPS drives destroy the data stored on them. A biometric drive that scans your fingerprint to provide access to its contents is another secure option.

✔ **9. Use SecurDisc technology to burn your files to disc.** The software maker Nero created SecurDisc with drive developer Hitachi-LG Data Storage. Through a combination of hardware and software, the SecurDisc system lets you use passwords to protect the CDs and DVDs you burn. It also provides digital signature verification, copy protection for PDFs, and a Data Integrity Check feature that alerts you if data on a disc has deteriorated. In addition, it uses sophisticated data verification tools that make it easier to recover data from damaged discs that were burned in a SecurDisc drive. To burn discs with SecurDisc technology, you must use a drive and software that support it. Companies that make SecurDisc-compatible drives include LG and Lite-On. Look for SecurDisc support in Nero's programs. You can find out more about the SecurDisc system at www.securdisc.net.

✔ **10. Use a USB flash drive with an application platform.** Flash drives with application platforms such as U3 (software.u3.com) and Ceedo (www.ceedo.com) can run software directly from the drive. That means you can plug one into any computer and open and edit files, without installing software or leaving any files on the computer. You can add security software for protecting files on the drive, and programs for viewing, playing, and editing photos, documents, videos, and music.

Cataloging Valuables & Personal Collections

Whether it's for insurance purposes, to sell items online, or simply to give yourself an organized perspective on your personal collections, creating a digital media catalog of your valuables can help you manage, preserve, and present information about them. Both digital photographs and video can be used to keep a record of your valuables. For some people, simply taking the images and storing them in a "Valuables" folder and in one remote location will be sufficient. If you want to use a software program to organize and annotate images of your valuables, and combine them with personal documents, try Microsoft's OneNote on a Windows system or Reinvented Software's Together (reinventedsoftware.com).

If you really want to go to town with your organizing or are a serious collector, you can purchase a dedicated home inventory program. This type of software allows you to add images and extensive text information to a database of your possessions or specific types of collections. You can also use optical character recognition (OCR) software to convert images of objects with text on them into searchable text files.

Use the Valuables Checklist to help ensure that you're not leaving any important items out of your catalog. If you're concerned about property loss from a possibly impending disaster (if you live in an area where hurricanes, wildfires, or other natural disasters are frequent, for example), or you have some other reason to need a record of an item quickly, check the Urgent box to help prioritize your documentation tasks. As you document each item, check the Documented box.

Ten Tips on Creating a Catalog of Your Valuables

✔ **1. Check your camera settings.** Use the highest resolution and quality settings available on your digital camera, and make sure the date and time are set correctly on it. That information will be saved in your photo metadata and might be helpful if you have to make an insurance claim. Pick a white balance setting that makes colors look natural under the light source you're using. When capturing the color of an item is critical, use manual white balance if your camera offers it. You will need to take a reference image of a white surface, such as a piece of paper, to set the manual white balance. If your photos come out blurry, use a tripod or turn up the lights so that the camera can use a higher shutter speed to reduce blur.

✔ **2. Get the lighting right.** Don't use a flash that will create a hot spot in your image, making details hard to see. If you need more light, use a diffused source such as a lamp with a shade, or bounce light onto the object you're photographing with a piece of white paper or posterboard. If you have a hard time capturing shiny objects such as coins without highlights and reflections that obscure details, try using a polarizing filter. If you don't have one or you can't attach one to your camera lens, try holding a polarized sunglass lens in front of your camera's lens to cut the glare. You may need to focus manually for this to work.

✔ **3. Shoot a video tour.** If you have a camcorder, you can use it to record all of your possessions by just walking around and shooting everything. Stop and zoom in on relevant details, and make descriptive comments. Turn the camcorder around to show yourself for a few minutes too. Make a copy of the video and put it in a safe deposit box or another secure location. If you have to make an insurance claim, you'll have strong proof that the items you lost were once in your home, with you. It's much harder to fake videos than it is photos.

✔ **4. Capture relevant details.** Zoom in and take shots of important details such as labels, model numbers, serial numbers, engravings, and stamps in jewelry and silverware. If you can't get close enough, use a macro or close-up mode. You can buy a close-up adapter lens for some cameras.

✔ **5. Take photos with voice annotations.** Some digital cameras allow you to make brief voice recordings and attach them to photos. That can be an efficient way to add information to your shots. Use image management software that can play back the voice annotations when you view the photos.

✔ **6. Add metadata.** You can use photo management software to add information about each item you photograph to the image file's metadata. Use a description, caption, or comment field for this. By adding metadata to the image file instead of entering it in a home inventory database, you'll ensure that the information stays with the photo if it is moved around and opened with other software or in an online gallery that supports metadata. Read more about photo metadata on page 84.

✔ **7. Show context.** If you're photographing items for insurance purposes, take some shots that show each object in the context of your home, instead of zooming in to crop out surrounding areas or setting up a plain white background. Showing items in the setting of your home may help you to prove ownership in the event of a loss. If you want to cre-

Software Resources ▸▸

OCR Software Makers
ABBYY www.abbyy.com
I.R.I.S. www.irislink.com
Nuance www.nuance.com
SimpleOCR www.simpleocr.com

Home Inventory Software Makers
Collectorz www.collectorz.com
Custom Apps www.cya2day.com
Frostbow Software frostbow.com
Kaizen Software Solutions
www.kzsoftware.com
Liberty Street Software
www.libertystreet.com
Mycroft Computing
www.mycroftcomputing.com
PrimaSoft PC www.primasoft.com
TurboSystemsCo
www.turbosystems.com
WenSoftware
www.wensoftware.com

ate more polished-looking versions of the photos later for online sales or other presentations, you can use masking software to remove the home background and substitute a professional-looking photo background. Vertus Bling!It (www.blingit.us) is an affordable and easy-to-use masking software option.

✔ **8. Show scale ▼.** If the size of the object you're photographing is important to its value, include a ruler or some other standard-sized object in the shot to show scale.

✔ **9. Don't edit your photos.** If you need your photos for insurance purposes, don't make any changes to them after you shoot them. In the event of a loss, you don't want to give your insurance company the impression that the images have been doctored in any way. Get the exposure and color settings right with your camera and lighting, so that adjustments to the images are not necessary. Make copies of your image files if you want to touch the photos up for online sales and other presentations.

✔ **10. Use data verification or authentication.** If you have very valuable items (meaning things that cost thousands of dollars, not hundreds), you might find it worthwhile to purchase a data verification or authentication kit. These work with only a small number of digital SLR cameras, so if you don't already own a compatible camera, that will be an additional expense. Canon makes a data verification kit that includes a special memory card and card reader in addition to software. Nikon provides a software-only image authentication kit. The technology is able to analyze images and determine whether they have been altered in any way, and produces results that are meant to hold up in court. Pricing for these products starts around $450.

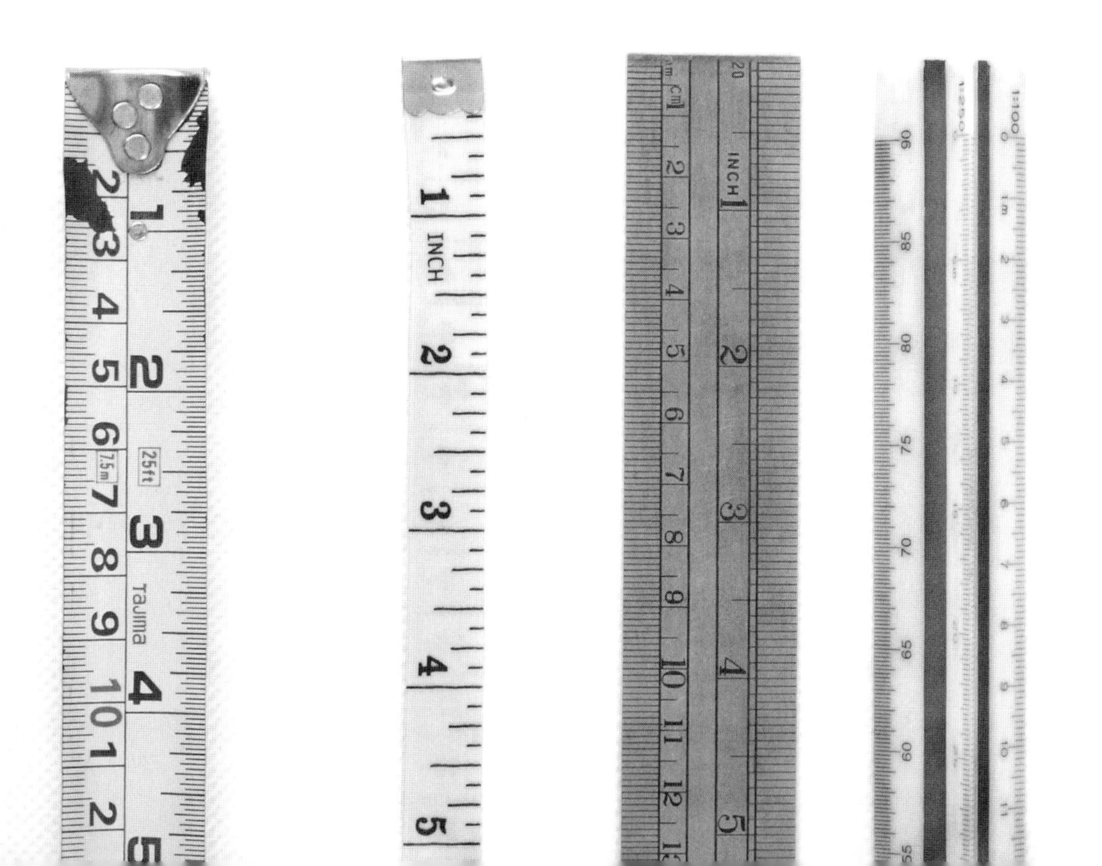

▶▶ Valuables Checklist

ITEMS	URGENT	DOCUMENTED
Antiques		
Art works		
Audio equipment		
Bicycles		
Boats		
Books		
Cars		
Camcorder equipment		
Camera equipment		
Collectibles		
Computers and peripherals		
Coins and medals		
Furniture		
Jewelry		
Kitchen appliances		
Memorabilia		
Misc. electronics		
Musical instruments		
Power tools		
Silverware		
Stamps		
TVs		
Watches		

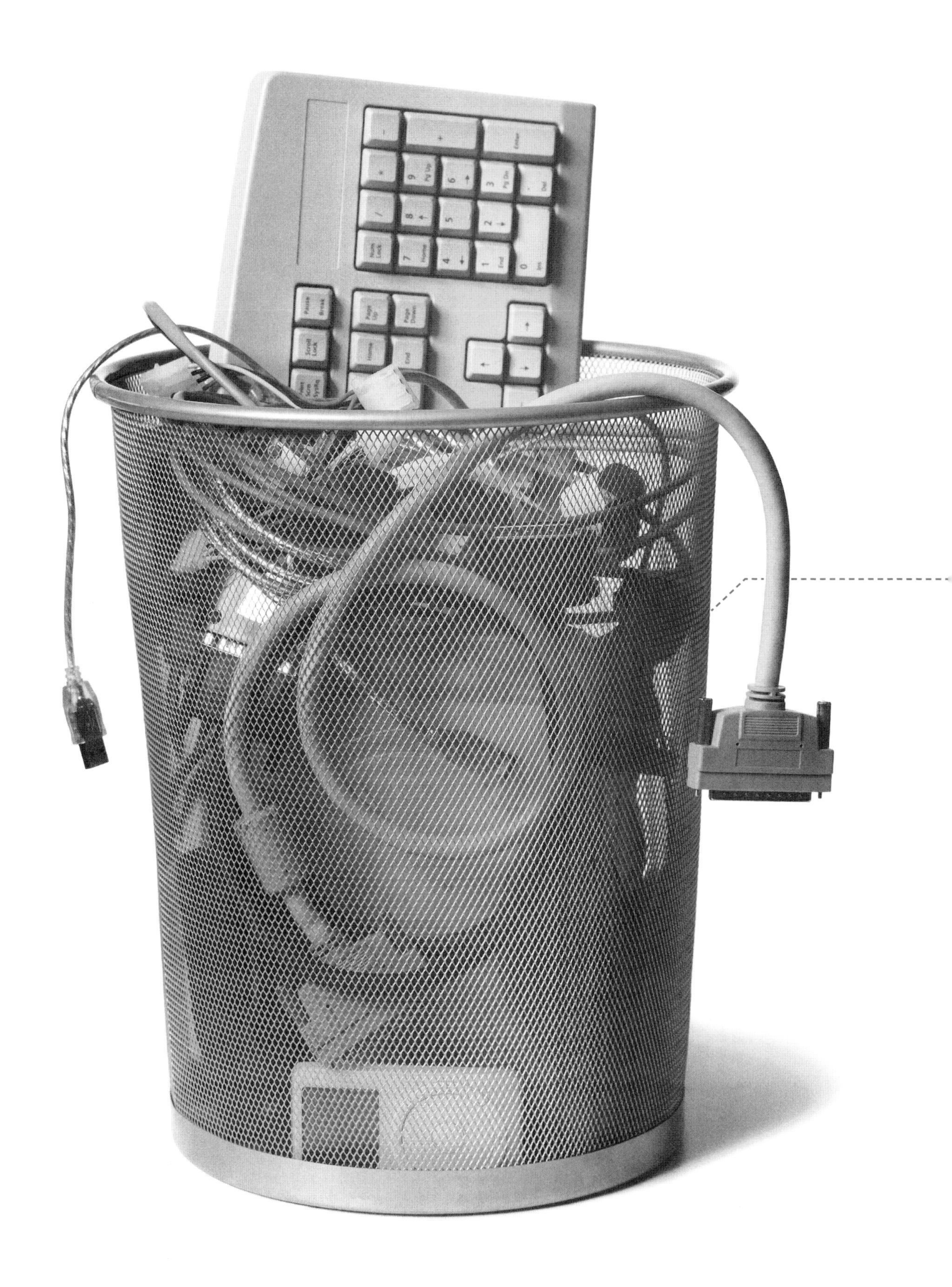

CHAPTER VI

Breaking Camp

Introduction:

Like all good homes, the shelter where you store your digital media will have to come down one day. It might be because you need to move your collection to a bigger place, or just because the old one is too rickety and in need of repair to be worth keeping. Given the pace at which technology moves these days, you'll probably face the process of upgrading your computer, storage devices, and media-playing hardware multiple times.

If you're assuming you should just delete your files from your old machines after you've copied them to new equipment, and then put the old gear out on the curb, think again. Not only is that an unwise choice in the age of data and identity theft, but it's the first step to transforming your old hardware into a dangerous pollutant.

In this chapter, you'll learn how to really, truly delete your files before you part with the devices or discs they're stored on, how to get rid of your old computer hardware and electronics without polluting the environment, and how to take your old hard drives out of broken machines so that you can retrieve or dispose of the files stored on them. You'll also get some tips on what to consider when you're leaving media to others or donating copies of work that you've created.

Deleting and Destroying Digital Files

Whether you recycle, resell, or give away your old computers and other devices, transferring them to another person can be like handing over a big box of your most important, private papers, photographs, and other media. Tossing old CDs, DVDs, and flash memory devices can amount to the same thing. Even if you carefully delete your files before getting rid of a device with a hard drive or flash memory, anyone with a little technical savvy and the right software could retrieve them. A few precautionary steps can help.

When you delete digital files, they are not actually removed from your system or device. What happens is more like a digital equivalent of clipping all the labels off of a drawer of file folders. The stuff is still in there, but your computer or device can't find it for you anymore. Reformatting a drive makes it even more difficult for the system to find the files, because reformatting destroys even more of the organizational structure used to locate and retrieve them. Still, a sophisticated data recovery program will probably be able to recover them.

Deleting files doesn't just remove their "labels," however, as in the file-cabinet analogy. It also gives your system permission to overwrite the deleted files, so that it can use the space they once took to store new files. If you keep adding data to a storage device, the files you have deleted will eventually be overwritten. For that reason, the way to clean out selected files (or all of the files) in your system is to use special software that overwrites them with random digital data and zeros. Such a program is usually called drive wiping or erasing software. (See the Resources list for a few recommended programs and where to buy them.) To ensure that it has deleted all of the data from your system or device before you relinquish it, make sure to run the software several times. Some programs can also be run on flash memory drives and cards, as well as rewritable optical discs such as CD-RWs and DVD-RWs.

If you're getting rid of a computer because it no longer works, however, you may not be able to run a drive wiping program. If your system doesn't function, but has a hard drive that may, you can remove the drive, then attach it to another computer—either by installing it in the other computer (see page 22 for instructions on installing a hard drive), or by using an external drive enclosure to connect it to another computer via USB. Once you've connected the drive, you can transfer its contents to the other computer, then run drive wiping software on it before disposing of the drive. That's a good option if the drive is old and you think it may be unreliable.

Disc Destroyer Makers ▸▸
Aleratec www.aleratec.com
Brando Workshop usb.brando.com.hk
Norazza norazza.com
Plextor www.plextor.com

You can purchase an inexpensive USB external hard drive enclosure for both desktop and laptop hard drives. Make sure you buy an enclosure that is made for the size and type of hard drive you have. Newer hard drives are usually Serial Advanced Technology Attachment (SATA) drives, while older ones are usually ATA drives, sometimes also called IDE, PATA, or EIDE.

If you're tossing old optical discs such as CDs or DVDs, you can render them unreadable by running them through a device that damages the layers on which data are stored. There are also machines that can shred discs, along with other thick items such as credit cards.

Here is how to remove your hard drive and install it in an external enclosure:

✔ **1. Make sure that you are not carrying a static charge** when you touch anything inside your computer. Touch something else that's metal before you start, to discharge any static electricity. Don't work near anything that might give you a charge, such as carpeting under your fuzzy slippers. You can buy an antistatic wristband from an electronics store and wear it while you work if your environment is static-prone.

✔ **2. Turn your computer off,** switch the power switch on the back of the machine off, and unplug it. If you're dealing with a laptop, remove the battery.

✔ **3. Remove the side panel** of your desktop computer so that you have access to the inside. You will probably have to unscrew the pegs that are holding it in place on the back of the computer case. If you're removing a hard drive from a laptop, you'll probably need to unscrew the cover of the hard drive compartment. Laptop designs vary, so if you're not sure how to get to the hard drive in yours, consult the manufacturer or the manual.

✔ **4. Look in the front** of your desktop computer and locate the hard drive. Unscrew and pull out the hard drive enclosure if necessary. Remove any screws holding the hard drive in place. Disconnect the cables or connectors attached to the hard drive.

✔ **5. Open the external drive enclosure** and connect the power and data connectors on your hard drive to the ones inside the enclosure.

✔ **6. Close the external enclosure** and plug it into your working computer's USB port. The hard drive should appear as an external drive on your computer.

✔ **7. Once the hard drive is attached** to your second, functional computer, you can transfer files from the old drive to your computer's other storage. Then run drive wiping software on the old drive before you dispose of it. Just be careful when you're selecting the drive to wipe so that you don't run the software on your computer's drive.

Getting Rid of Hardware

It doesn't require mathematical genius to reach the conclusion that all the outdated computers and electronics you've discarded over the years, multiplied by the number of people who have been doing the same thing, equals a whole lot of high-tech garbage. Technology moves fast, which means that we're constantly upgrading our computers, phones, cameras, and other gizmos—and getting rid of older devices.

For the most part, we're still sending our old gear straight to the landfill. And even when we recycle, the job isn't always done right. Instead of refurbishing old devices or using safe processes to recycle their materials, some companies simply sell the items they collect to a broker in a developing country. What happens after that often isn't pretty. Reports from a range of sources have shown unprotected workers, sometimes children, being exposed to highly toxic materials as they burn and disassemble the devices to extract valuable material such as metals.

Whether your old machines end up in that situation or in a landfill, the toxic materials they contain find their way into the ground and waterways, and possibly into new commercial products made by manufacturers who purchase salvaged metals to use as cheap raw materials. A recent study by scientists at Ashland University tested pieces of inexpensive jewelry sold in the United States, and concluded that the toxic lead they contained probably came from circuit board solders. Research aside, common sense dictates that pouring toxic material into the ground and water isn't the best way to ensure the safety of our food supply.

There are many ways to dispose of your old computer and electronic hardware safely, and sometimes even profitably. When it comes time to upgrade—or to get rid of analog devices after digitizing your older media—don't just put your old stuff on the curb or hand it over to a recycler that you haven't checked out. It's easy to find recyclers that have been approved by monitoring organizations (see the Resources list).

Many manufacturers are making efforts these days to reduce the amounts of toxic materials that are used in computers and electronics in the first place, and to make them more energy-efficient. So when you upgrade, look for products that take less of a toll on the environment—and your wallet—when you use them and when it's time to part with them.

Ten Ways to Get Rid of Hardware and Buy Green New Gear

✔ **1. Donate your hardware.** If you're upgrading, but your older equipment still works, consider giving it to a school, a club, or another organization that takes donations. You may be able to write off the donation on your taxes. Make sure to wipe the memory of any device you give away. Read about how to do that on page 174.

✔ **2. Recycle your hardware.** Many manufacturers have takeback programs that allow you to send in old gear for recycling, sometimes in exchange for a discount on new items. Another way to have your hardware recycled is to take it to a retailer that offers a recycling program. Many office supply and electronics stores do, along with some groceries and other types of stores. Some charge a small fee for accepting larger items such as desktop computers; if you can't find a free program to take your hardware, consider the fee of $10 or so to be your donation toward a good environmental cause. Many of the sites in the Resources list in this section provide online listings of local and mail-in recycling programs. Make sure that you pick a recycler that follows ethical, environmentally sound practices, and doesn't simply sell the hardware it collects to an overseas broker. A good place to start looking for an ethical recycler is the Basel Action Network's listing of eSteward-certified recyclers.

✔ **3. Sell your hardware.** Making a few dollars from your older hardware is always an option if it's still in good working condition. As with donations, you should make sure to wipe the memory of any gear you put on the block. Consider selling your gear on Web sites such as Craigslist (www.craigslist.org) or eBay (www.ebay.com), or through an ad posted in a local outlet such as a coffee shop or a school or work bulletin board.

Recycling Resources ▸▸

These organizations provide a wealth of information about reliable recyclers, manufacturers that have takeback programs, groups that accept donations, and the ways in which e-waste affects the environment and human health.

Basel Action Network www.ban.org
Computer TakeBack Campaign www.computertakeback.com
CTIA - The Wireless Association recyclewirelessphones.com
Earth 911 earth911.org
EMPA E-waste Guide ewasteguide.info
GreenCitizen www.greencitizen.com
Greenpeace www.greenpeace.org
National Cristina Foundation www.cristina.org
National Geographic Society www.nationalgeographic.com
National Resources Defense Council www.nrdc.org
Phonefund www.phonefund.com
Rechargeable Battery Recycling Corporation call2recycle.org
StEP (Solving the E-waste Problem) www.step-initiative.org
Telecommunications Industry Association, E-cycling Central www.eiae.org
The Wireless Foundation www.wirelessfoundation.org
United States Environmental Protection Agency www.epa.gov

✔ **4. Use the U.S. Postal Service Mail Back program.** The United States Postal Service (www.usps.gov) offers a free recycling service for small electronics and printer cartridges in some areas. The program makes free, postage-paid Mail Back envelopes available in post office lobbies. To recycle approved items, you just put them in one of the envelopes and drop it in the mail box. Check with your local post office to find out if the Mail Back program is offered in your area.

✔ **5. Reuse newer components.** If you have upgraded your computer's components at some point, or if you paid extra for certain premium components when you bought the machine, you might want to keep them when you get rid of the rest of the computer. Installing them in your new computer instead of paying for brand-new high-end components may save you some money, too. This can be a good option for people who are comfortable enough with technology to remove and install components—or who know someone who is. Just make sure that the components are compatible with your new machine before you start tinkering.

✔ **6. Use a disposal service.** You can read about how to wipe the data from the memory of your computer or other device before getting rid of it on page 174. However, if you don't want to handle that process yourself, or if you want to be absolutely sure that no data can ever be extracted from your old hardware, you can send your hard drive to a service that will physically destroy it for a fee. Some services can even handle whole computers and electronic devices. Choose one that follows environmentally responsible procedures for recycling the material that is left after the destruction process.

✔ **7. Recycle the small stuff.** Batteries and ink cartridges ▸ contain toxic pollutants too, so don't just toss them in the trash when they're used up. Some printer makers have takeback programs that allow you to mail your used cartridges in or drop them off at a store, in exchange for a discount on new cartridges. Recycling programs that accept electronics sometimes accept batteries as well. For convenience, you can keep a jar or box for used batteries and cartridges under your kitchen sink or in the back of a desk drawer. Recycle when you fill it up.

✔ **8. Buy EnergyStar products.** When you replace your old computers and electronics, look for new ones that meet the EnergyStar criteria. EnergyStar (www.energystar.gov) is a program of the U.S. Environmental Protection Agency and Department of Energy that promotes energy-efficient products. Computers and electronics that carry the EnergyStar label are more energy-efficient than comparable competitors. That's better for both your electricity bill and the environment.

✔ **9. Check the EPEAT registry.** The Electronic Product Environmental Assessment Tool (EPEAT) is a system that helps consumers compare computers and monitors according to how environmentally friendly they are. It uses a list of fifty-one criteria, and assigns simple gold, silver, and bronze ratings to approved products. You can check the EPEAT product registry at www.epeat.net before you buy a new computer or monitor to see how it rates. Some retailers, such as Buy.com, include EPEAT ratings in the product information they provide to shoppers.

✔ **10. Buy RoHS-compliant products.** The European Union requires that computers and electronics sold in its member countries conform to restrictions on levels of lead, cadmium, mercury, hexavalent chromium, polybrominated biphenyl (PBB), and polybrominated diphenyl ether (PBDE). Products that comply with this restriction on the use of hazardous substances are often labeled as RoHS-compliant when sold in the U.S. and elsewhere outside the European Union. You can read more about it at www.rohs.eu.

How to Retain Copyright When Giving Media Away

When you create photographs, music, video, and some other types of work, copyright is granted to you automatically when you complete the work. You don't have to register copyright of your own work in order to own it. However, registering is a good idea if you are making copies of the work available to others, including organizations, and want to ensure that your copyright ownership (or your heirs') is clear and easy to prove should there be any confusion or legal dispute about it in the future. If the items in question were not created by you, but were inherited or obtained in some other way, you do not own their copyright unless it has been legally granted to you. If there is any confusion about this issue, consult a lawyer.

You can register the digital media works that you have created yourself through the United States electronic Copyright Office (eCO), which is part of the United States Copyright Office (www.copyright.gov). The online service allows you to register multiple digital works at once, by uploading them, filling an online form, and paying an affordable fee. Media created or copyrighted by different parties must be registered separately. The Copyright Office Web site also provides instructions for registering works by mail, as well as extensive information about copyright laws and procedures.

Don't forget to include these items in your will if it's important to you that they be preserved or inherited by particular people. If you want your heirs to inherit the copyright for the media you created, you should specify that in your will. Also ensure that the media you want to leave to others will be identifiable and accessible to the people that you want to pass it on to. Include any passwords or other security information in your will or in other documents that are stored in a safe deposit box or another location that is secure.

Will Resources ▸▸
These sites provide information on how to write a legal will, as well as online tools and software that can help you do it.

American Bar Association www.abanet.org
BuildaWill www.buildawill.com
LegacyWriter www.legacywriter.com
LegalZoom www.legalzoom.com
Nolo www.nolo.com
Quicken Intuit WillMaker quicken.intuit.com

Resources

Here is a full list of all the resources mentioned in this book in order of topic. Some lists were truncated in the main text for space reasons. All of them have been vetted and approved by the author.

Multimedia Organizers
ArcSoft MediaImpression www.arcsoft.com
J. River Media Center www.jrmediacenter.com
Microsoft Expression Media
www.microsoft.com/expression

Hard Drive, Enclosure, and NAS makers
Apple www.apple.com
Beyond Micro www.beyondmicro.com
Buffalo Technology www.buffalotech.com
Data Robotics www.drobo.com
D-Link www.dlink.com
Fantom Drives www.fantomdrives.com
Iomega www.iomega.com
Kanguru Solutions www.kanguru.com
LaCie www.lacie.com
Netgear www.netgear.com
Synology www.synology.com
Western Digital www.westerndigital.com

Flash Memory Card & USB drive makers
ATP www.atpinc.com
Kingston www.kingston.com
Lexar www.lexar.com
PNY www.pny.com
SanDisk www.sandisk.com
SimpleTech www.simpletech.com

Optical Disc Information
Blu-ray.com www.blu-ray.com
Blu-ray Disc Association www.blu-raydisc.com
CDFreaks www.cdfreaks.com
CDRLabs www.cdrlabs.com
DVD Forum www.dvdforum.org
Optical Storage Technology Association www.osta.org

Disc Manufacturers
Maxell www.maxell.com
Mitsui MAM-A www.mam-a.com
Taiyo Yuden www.t-yuden.com
Verbatim www.verbatim.com

Disc-Burning Software Makers
Nero www.nero.com
Padus DiscJuggler www.padus.com
Roxio www.roxio.com

Disc and Drive Utility Software
DVDInfoPro www.dvdinfopro.com
Nero DiscSpeed www.cdspeed2000.com
Plextor PlexTools www.plextools.com

Backup and Sync Software Makers
Acronis www.acronis.com
Centered Systems www.centered.com
EMC Insignia www.emcinsignia.com
Genie-Soft www.genie-soft.com
Host Interface International www.hostinterface.com
Macrium www.macrium.com
Memeo www.memeo.com
Nero www.nero.com
NewTech Infosystems (NTI) www.ntius.com
NovaStor www.novastor.com
Novosoft www.handybackup.net
Paragon Software Group
www.paragon-software.com
Prosoft Engineering www.prosofteng.com
Roxiowww.roxio.com
Siber Systems www.goodsync.com
Softland www.backup4all.com
StorageCraft Technology www.storagecraft.com
Symantec www.symantec.com
TGRMN Software www.tgrmn.com
Titan www.titanbackup.com

Backup and Sync Services
Apple MobileMe www.apple.com/mobileme
4Shared www.4shared.com
backup.com backup.com
Bluestring www.bluestring.com
Box.net box.net
Carbonite www.carbonite.com
DropBoks www.dropboks.com
Fabrik www.fabrik.com
File Den www.fileden.com
FlipDrive www.flipdrive.com
IBackup www.ibackup.com
IDrive www.idrive.com

Intronis eSureIT Home www.intronis.com
Iomega iStorage www.iomega.com
Iron Mountain backup.ironmountain.com
Jungle Disk www.jungledisk.com
Laplink www.laplink.com
Memeo Share www.memeo.com
Microsoft Windows Live home.live.com
Mozy mozy.com
Openomy www.openomy.com
Ovi ovi.nokia.com
Proxure KeepVault www.keepvault.com
SugarSync www.sugarsync.com
Transmedia Glide www.transmediacorp.com
Xdrive www.xdrive.com

Remote Access Tools

CyberLink Live www.cyberlinklive.com
GoToMyPC www.gotomypc.com
Magnetk ExpanDrive www.expandrive.com
Magnetk SftpDrive www.sftpdrive.com
MioNet www.mionet.com

Peer-to-Peer Online Storage

Wuala wua.la

Independent Product Testers

Intertek www.intertek-etlsemko.com
Underwriters Laboratories www.ul.com

Safe Manufacturers

Cobalt Safes www.cobaltsafes.com
Fire Fyter www.firefyter.com
FireKing Security Group www.fireking.com
Honeywell Safes www.honeywellsafes.com
SentrySafe www.sentrysafe.com

Waterproof Case Makers

OtterBox www.otterbox.com
Pelican www.pelican.com
Seahorse www.seahorse.net
Underwater Kinetics www.uwkinetics.com

Memory Card Recovery Software

Best IT Solutions PhotoOne Recovery www.photoone.net
DataRescue www.datarescue.com
Disk Doctors Digital Media Recovery www.diskdoctors.net
Galaxy Digital Photo Recovery www.photosrecovery.com
Image Recall Don't Panic www.imagerecall.com
MediaRECOVER www.mediarecover.com

Hard Drive Recovery Software

Chily Softech RecoveryFIX www.recoveryfix.com
Get Data Recover My Files www.recovermyfiles.com
Salvage Data SalvageMedia www.salvagedata.com

Optical Disc Recovery Software

InfinaDyne www.cdrom-prod.com
Octane Soft inDisc Recovery ww.octanesoft.com
Salvage Data SalvageMedia www.salvagedata.com
Smart Projects IsoBuster www.smart-projects.net

Data Recovery Services

Acodisc www.acodisc.com
Aero Data Recovery www.aerodr.com
CD Data Guys www.cddataguys.com
DriveSavers Data Recovery www.drivesavers.com
DTI Data Recovery www.dtidata.com
First Advantage Data Recovery Services www.datarecovery.net
Nucleus Data Recovery www.nucleusdatarecovery.org
Ontrack Data Recovery www.ontrackdatarecovery.com
Salvage Data Recovery Service www.salvagedata.com
Seagate Recovery Services services.seagate.com
Stellar Information Systems www.stellarinfo.com

Hard Drive Diagnostic Software

Ariolic Active SMART www.ariolic.com
PassMark DiskCheckup www.passmark.com
Stellar Information Systems Stellar Smart www.stellarinfo.com

Disc Resurfacing Device Makers

Aleratec www.aleratec.com
Digital Innovations www.digitalinnovations.com

Film Scanner Manufacturers

Braun www.braun-phototechnik.de
Nikon www.nikon.com
Pacific Image www.scanace.com

Plustek www.plustek.com

Flatbed Scanner Manufacturers

Canon www.canon.com
Epson www.epson.com
Hewlett-Packard www.hp.com
Microtek www.microtek.com
Umax www.umax.com

Scanning Software

Hamrick Software VueScan www.hamrick.com
LaserSoft Imaging SilverFast www.silverfast.com

Scanner Reviews

CNET www.cnet.com
Imaging Resource www.imaging-resource.com
PC Magazine www.pcmag.com
Popular Photography & Imaging www.popphoto.com

Image Conversion Programs

fCoder Group Image Converter Plus www.imageconverterplus.com
Mystik Media AutoImager www.autoimager.com
Online-Utility Image Converter www.online-utility.org/image_converter.jsp
ReaSoft ReaConverter www.reasoft.com

Scanning Services

BritePix www.britepix.com
Digital Memories digitalmemoriesonline.net
Digital Pickle www.digitalpickle.com
DigMyPics www.digmypics.com
FotoBridge www.fotobridge.com
Gemega Imaging www.gemega.com
Larsen Digital Services www.slidescanning.com
My Special Photos www.myspecialphotos.com/
Pixmonix www.pixmonix.com
ScanCafe www.scancafe.com
ScanMyPhotos www.scanmyphotos.com

Free Photo Editing and Management Tools

Apple iPhoto (included in the iLife suite that usually comes with a Mac, but can also be purchased separately) www.apple.com
Google Picasa picasa.google.com
Kodak EasyShare www.kodak.com
Windows Vista: included Explorer & Photo Gallery www.microsoft.com
Windows XP: included Explorer and downloadable Live Photo Gallery www.microsoft.com

Image Editing Software with Management Tools

Adobe Photoshop Elements with Adobe Bridge www.adobe.com
Arcsoft PhotoImpression www.arcsoft.com
Corel MediaOne www.corel.com
FotoFinish www.fotofinish.com
Nova Development Photo Explosion Deluxe www.novadevelopment.com
Roxio PhotoSuite www.roxio.com

Advanced Photo Workflow Software

ACDSee 10 Photo Manager www.acdsee.com
Adobe Lightroom (RAW image workflow software) www.adobe.com
Adobe Photoshop CS3 with Adobe Bridge www.adobe.com
Apple Aperture (RAW image workflow software) www.apple.com
Bibble Labs Bibble Pro (RAW image workflow software) www.bibblelabs.com
Light Crafts LightZone www.lightcrafts.com
Phase One Capture One (RAW image workflow software) www.phaseone.com
Corel Paint Shop Pro X2 www.corel.com

Advanced Photo Management Software

ACDSee Pro 2 Photo Manager www.acdsee.com
Arcsoft MediaImpression www.arcsoft.com
Breeze Systems BreezeBrowser Pro www.breezesys.com
Camera Bits Photo Mechanic www.camerabits.com
Cerious Software ThumbsPlus www.cerious.com
Extensis Portfolio www.extensis.com
FastStone Image Viewer www.faststone.org
FotoTime FotoAlbum www.fototime.com
IDimager www.idimager.com
IrfanView www.irfanview.com
Microsoft Expression Media www.microsoft.com
Photools IMatch www.photools.com
PicaJet www.picajet.com
Pro Shooters DigitalPro www.proshooters.com
Xequte Smart Pix Manager www.xequte.com

Geocoding Loggers and Cameras with Built-in GPS

ATP Electronics www.atpinc.com
Dawn Technology www.dawntech.hk
General Electronics www.general-imaging.com
GiSTEQ www.gisteq.com
Jobo www.jobo.com
Nikon www.nikon.com
Pharos www.pharosgps.com
Red Hen Systems www.redhensystems.com
Sony www.sony.com

Automatic Geocoding Software

Atomix Technologies JetPhoto Studio www.jetphotosoft.com
Breeze Systems Downloader Pro www.breezesys.com
GeoSpatial Experts www.geospatialexperts.com
Houdah Software Houdahgeo www.houdah.com
locr GPS Photo www.locr.com
MMI Software PhotoGPSEditor www.mmisoftware.co.uk
Ovolab Geophoto www.ovolabs.com
OziPhotoTool www.oziphototool.com
Pretek RoboGEO www.robogeo.com
World Wide Media Exchange Location Stamper http://wwmx.org

Camera Phone Geocoding Software

GEOsnapper Mobile www.geosnapper.com
locr GPS Photo www.locr.com
Merkitys-Meaning http://meaning.3xi.org
ShoZu Share-it www.shozu.com
viewranger www.viewranger.com
ZoneTag http://zonetag.research.yahoo.com

Manual Geocoding Software

Arcsoft MediaImpression www.arcsoft.com
Google Picasa http://picasa.google.com

Online Photo Gallery and Sharing Sites That Support Geocoded Photos

GEOsnapper www.geosnapper.com
Google Picasa Web Albums http://picasa.google.com
ipernity www.ipernity.com
locr www.locr.com
Google Panoramio http://www.panoramio.com
pikeo www.pikeo.com
SmugMug www.smugmug.com
Klika TripTracker http://triptracker.net
Woophy http://woophy.com
Yahoo! Flickr www.flickr.com
Zooomr www.zooomr.com

Mobile/Desktop Synchronization Software Makers

FutureDial www.futuredial.com
Mobile Action global.mobileaction.com
MobTime www.mobtime.com
Susteen www.susteen.com

Photo Printer Manufacturers

Canon www.canon.com
Epson www.epson.com
Hi-Touch Imaging www.hitouchimaging.com
Hewlett Packard www.hp.com
Kodak www.kodak.com
Sony www.sony.com

Printing Software

Arcsoft Print Creations and PhotoPrinter www.arcsoft.com
Canon Easy-PhotoPrint www.canon.com (look under Canon printer model names in the Support section)
Qimage www.ddisoftware.com/qimage

Monitor Calibrator Makers

ColorVision www.datacolor.com
X-Rite www.xrite.com

Printer Review Sources

CNET www.cnet.com
Imaging Resource www.imaging-resource.com
PC Magazine www.pcmag.com
PC World www.pcworld.com
Popular Photography & Imaging www.popphoto.com
Printer Info www.printerinfo.com

Archival Supply Retailers

Archival Methods www.archivalmethods.com
Conservation Resources International www.conservationresources.com
Hollinger Corporation www.hollingercorp.com
Light Impressions www.lightimpressionsdirect.com
Talas talasonline.com
University Products Archival Suppliers www.archivalsuppliers.com

Digital Photo Frames

PhotoShowTV www.simplestar.com
Ceiva Digital Frame www.ceiva.com)
WiFi Digital Frame www.estarling.com

Gallery Sites for Photo Archiving

Flickr www.flickr.com
Fotki www.fotki.com
ImageEvent imageevent.com
Kodak Gallery www.kodakgallery.com
Phanfare www.phanfare.com
Shutterfly www.shutterfly.com
SmugMug www.smugmug.com
Snapfish www.snapfish.com
Swiss Picture Bank www.swisspicturebank.com
Winkflash www.winkflash.com

Online Gallery & Photo Community Sites

Adobe Photoshop Express www.adobe.com/products/photoshopexpress
AOL Pictures pictures.aol.com
dotPhoto www.dotphoto.com
DropShots www.dropshots.com
fotocommunity www.fotocommunity.com
Fujifilm.net www.fujifilm.net
GEOsnapper www.geosnapper.com
Google Picasa Web Albums picasaweb.google.com
ipernity www.ipernity.com
Jalbum jalbum.net
JPG Magazine www.jpgmag.com
KoffeePhoto www.koffeephoto.com
locr www.locr.com
Nikon my Picturetown www.mypicturetown.com
Panoramio www.panoramio.com
PBase www.pbase.com
Photo.net photo.net
Photobucket photobucket.com
Photocheap www.photocheap.biz
Photomax www.photomax.com
PhotoWorks www.photoworks.com
Picturetrail www.picturetrail.com
Piczo www.piczo.com
pikeo www.pikeo.com
Pixagogo www.pixagogo.com
Plazes www.plazes.com
Printroom www.printroom.com
Sacko www.sacko.com
TripTracker triptracker.net
Webshots www.webshots.com
Woophy www.woophy.com
Worldisround www.worldisround.com
ZoomIn www.zoom.in
Zooomr www.zooomr.com

Dual VCR/DVD deck makers

JVC www.jvc.com
Panasonic www.panasonic.com
Philips www.philips.com
Samsung www.samsung.com
Sony www.sony.com
Toshiba www.toshiba.com
Zenith www.zenith.com

Analog-to-Digital Conversion Device Makers

ADS Tech www.adstech.com
AMD www.amd.com
Belkin www.belkin.com
Pinnacle www.pinnaclesys.com
Plextor www.plextor.com
StarTech www.startech.com

Film Cleaning Solution Makers

ECCO
Edwal
FilmRenew
VitaFilm
Xekote
prositesclementsx.homestead.com

Media Players with Organizing Functions

Adobe Media Player get.adobe.com/amp
Apple iTunes www.apple.com
Microsoft Windows Media Player www.microsoft.com
Nullsoft Winamp www.winamp.com
RealNetworks RealPlayer www.real.com

Video Editing Software for the Casual User

Adobe Premiere Elements www.adobe.com
Apple Final Cut Express www.apple.com
Apple iMovie www.apple.com
Arcsoft VideoImpression www.arcsoft.com
CyberLink PowerDirector www.cyberlink.com
Microsoft Windows Movie Maker www.microsoft.com

Nero Vision (part of Nero Ultra Edition) www.nero.com
Pinnacle Studio and Studio Plus www.pinnaclesys.com
Roxio Easy Media Creator www.roxio.com
Sony Vegas Movie Studio www.sonycreativesoftware.com
Ulead VideoStudio www.ulead.com

Video Editing Software for Professionals

Adobe Premiere Pro www.adobe.com
Apple Final Cut Pro www.apple.com
Pinnacle Studio Ultimate www.pinnaclesys.com
Sony Vegas Pro www.sonycreativesoftware.com

Video Management Tools

Aquafadas iDive www.aquafadas.com
DVdate paul.glagla.free.fr/index_en.htm
ScenalyzerLive www.scenalyzer.com

Software to Catalog Commercial Movies

Collectorz Movie Collector www.collectorz.com
TurboSystemsCo Video Librarian Plus www.turbosystems.com
PrimaSoft Movie Organizer Deluxe www.primasoft.com

Online Video Hosting Sites

AtomUploads www.atomfilms.com
Blip.tv blip.tv
Break.com www.break.com
Buzznet www.buzznet.com
Dailymotion www.dailymotion.com
GameVideos www.gamevideos.com
Google Video video.google.com
imeem www.imeem.com
kewego www.kewego.com
Liveleak www.liveleak.com
Metacafe www.metacafe.com
MSN Soapbox video.msn.com
OneWorldTV tv.oneworld.net
Ourmedia www.ourmedia.org
Ovi share.ovi.com
pandora tv www.pandora.tv
Peekvid.com peekvid.com
Revver www.revver.com
sevenload en.sevenload.com
Veoh www.veoh.com
Vimeo www.vimeo.com
Vuze www.vuze.com
Yahoo! Video video.yahoo.com
YouAreTV www.youare.tv
YouTube www.youtube.com

Recording Software Makers

Acoustica www.acoustica.com
AIPL www.aipl.com
AlpineSoft www.alpinesoft.co.uk
Audacity audacity.sourceforge.net
Audiotool www.audiotool.net
AVSMedia www.avsmedia.com
Cakewalk www.cakewalk.com
CFB Software www.cfbsoftware.com
Channel D www.channld.com
Coyote Electronics www.coyotes.bc.ca
NCH Software www.nch.com.au
PolderbitS www.polderbits.com
Roxio www.roxio.com
TongSoft www.tongsoft.com
Tracer Technologies www.tracertek.com
Wieser Software www.ripvinyl.com

USB Turntable Makers

Audio-Technica www.audio-technica.com
Grace Digital Audio www.gracedigitalaudio.com
ION Audio www.ion-audio.com
Numark www.numark.com
TEAC America, Inc. www.teac.com

Media Player & Music Management Software

Apple iTunes www.apple.com
Duplicate Music Files Finder www.lcibrossolutions.com
Media Catalog Studio www.maniactools.com
Media Jukebox www.mediajukebox.com
MediaMonkey www.mediamonkey.com
Microsoft Windows Media Player www.microsoft.com
MP3tag www.mp3tag.de/en
Music Label www.codeaero.com
Music Library www.wensoftware.com/MusicLibrary
MusicBrainz Picard musicbrainz.org
Teen Spirit teenspirit.artificialspirit.com
The GodFather (from www.download.com)
TuneUp www.tuneupmedia.com
Winamp www.winamp.com

Digital Music Receiver, Transmitter, and Player Makers

Apple www.apple.com
Buffalo Technology www.buffalotech.com
Choice Select www.choiceselectonline.com
Denon www.denon.com
D-Link www.dlink.com
Freecom www.freecom.com
Hauppauge www.hauppauge.com
Linksys www.linksys.com
Linn www.linn.co.uk
Logitech www.logitech.com
Netgear www.netgear.com
Onkyo www.onkyo.com
Panasonic www.panasonic.com
Philips www.philips.com
Pioneer www.pioneerelectronics.com
Roku www.roku.com
Slim Devices www.slimdevices.com
Sonos www.sonos.com
Sony www.sony.com
X10 www.x10.com

CD Ripping Services

Moondog Digital www.moondogdigital.com
MusicShifter http://musicshifter.com
Pickled Productions www.pickledproductions.com
ReadyToPlay www.readytoplay.com
RipDigital www.ripdigital.com
Ripstyles www.ripstyles.com
Riptopia www.riptopia.com

Secure USB Flash Drive Makers

IronKey www.ironkey.com
Kanguru www.kanguru.com
Kingston Technology www.kingston.com
Lexar www.lexar.com
SanDisk www.sandisk.com

OCR Software Makers

ABBYY www.abbyy.com
I.R.I.S. www.irislink.com
Nuance www.nuance.com
SimpleOCR www.simpleocr.com

Home Inventory Software Makers

Collectorz www.collectorz.com
Custom Apps www.cya2day.com
Frostbow Software frostbow.com
Kaizen Software Solutions www.kzsoftware.com
Liberty Street Software www.libertystreet.com
Mycroft Computing www.mycroftcomputing.com
PrimaSoft PC www.primasoft.com
TurboSystemsCo www.turbosystems.com
WenSoftware www.wensoftware.com

Makers of Drive Wiping Software

Acronis www.acronis.com
CyberScrub www.cyberscrub.com
DBAN dban.sourceforge.net
DTI Data www.dtidata.com
Heidi Computers www.heidi.ie
Jetico www.jetico.com
LSoft Technologies www.killdisk.com
Migo Software www.migosoftware.com
Object Rescue www.objectrescue.com
Paragon Software Group
www.paragon-software.com
QSGI www.eraseyourharddrive.com
Rixstep rixstep.com
Stellar Information Systems www.drive-wipe.com
Webroot Software www.webroot.com
WhiteCanyon www.whitecanyon.com

Disc Destroyer and Shredder Makers

Aleratec www.aleratec.com
Brando Workshop usb.brando.com.hk
Norazza norazza.com
Plextor www.plextor.com

Recycling

Basel Action Network www.ban.org
Computer TakeBack Campaign
www.computertakeback.com
CTIA - The Wireless Association
recyclewirelessphones.com
Earth 911 earth911.org
EMPA E-waste Guide ewasteguide.info
GreenCitizen www.greencitizen.com
Greenpeace www.greenpeace.org
National Cristina Foundation www.cristina.org
National Geographic Society
www.nationalgeographic.com
National Resources Defense Council
www.nrdc.org
Phonefund www.phonefund.com

Rechargeable Battery Recycling Corporation
call2recycle.org
StEP (Solving the E-waste Problem)
www.step-initiative.org
Telecommunications Industry Association
E-cycling Central www.eiae.org
The Wireless Foundation www.wirelessfoundation.org
Environmental Protection Agency www.epa.gov

Will Resources

American Bar Association
www.abanet.org
BuildaWill www.buildawill.com
LegacyWriter www.legacywriter.com
LegalZoom www.legalzoom.com
Nolo www.nolo.com
Quicken Intuit WillMaker quicken.intuit.com

Acknowledgments

Putting this book together required a great deal of work by many people, and I extend my gratitude to all of them. Thanks especially to my editor Bronwen Latimer for providing the inspiration for the book and orchestrating its publication, to Esther Ferington for her careful attention to the text and helpful ideas, and to Melissa Farris for making the whole thing look great. I also appreciate the assistance of representatives of the many companies whose products I took for a spin during the course of writing the book.

Most of all, I'd like to thank my family and my indispensible friends for their love and encouragement. Very special thanks to Anne, Bryan, and Serhat. And thanks, as always, to my parents, for everything.

Illustration Credits

1, Tomasz Darul/Shutterstock; 2, Peter Doomen/Shutterstock; 6, Kuzma/Shutterstock; 8, Koksharov Dmitry/Shutterstock; 9, Jorge Salcedo/Shutterstock; 14, Bomshtein/Shutterstock; 16, Alexan66/Shutterstock; 17, Povl E. Petersen/Shutterstock; 18, Phaif/Shutterstock; 21, Rafa Irusta/Shutterstock; 31, Tony Mathews/Shutterstock; 32, luminouslens/Shutterstock; 35, 5607594264/Shutterstock; 44, James Steidl/Shutterstock; 45, Shapiso/Shutterstock; 46, Dic Liew/Shutterstock; 48, Milat/Shutterstock; 49 (UP), Bocos Benedict/Shutterstock; 49 (LO), pixelman/Shutterstock; 50, Franck Boston/Shutterstock; 53, Bocos Benedict/Shutterstock; 54, 8781118005/Shutterstock; 56, Paul Cowan/Shutterstock; 58, Michael Ransburg/Shutterstock; 66, Per-Anders Jansson/Shutterstock; 69, ozgur artug/Shutterstock; 71, sabri deniz kizil/Shutterstock; 73, Elena Pal/Shutterstock; 75, Tomasz Darul/Shutterstock; 82, skaljac/Shutterstock; 84, Luis Marden; 84 (Inset RT), NatUlrich/Shutterstock; 84 (Inset LE), fotorro/Shutterstock; 89, ozgur artug/Shutterstock; 91, sabri deniz kizil/Shutterstock; 95 (LO), Courtesy Nikon; 100, sabri deniz kizil/Shutterstock; 101, Jorge Enrique Villalobos/Shutterstock; 104, Graca Victoria/Shutterstock; 105, Pavel Drozda/Shutterstock; 108, 5607594264/Shutterstock; 114, James Steidl/Shutterstock; 115, Leo/Shutterstock; 116, Valery Potapova/Shutterstock; 120, Stefan Jovanovic/Shutterstock; 123 (UP), sabri deniz kizil/Shutterstock; 123 (CTR), sabri deniz kizil/Shutterstock; 123 (LO), sabri deniz kizil/Shutterstock; 125, ayazad/Shutterstock; 132, sabri deniz kizil/Shutterstock; 134, Nicholas Piccillo/Shutterstock; 136, Marko Poplasen/Shutterstock; 137, Arvind Balaraman/Shutterstock; 138, Mark Thiessen, NGS; 141, J. Helgason/Shutterstock; 142, c./Shutterstock; 145, R/Shutterstock; 146 (UP), emin kuliyev/Shutterstock; 146 (UP CTR), Joseph McCullar/Shutterstock; 146 (LO CTR), dubassy/Shutterstock; 146 (LO), 5607594264/Shutterstock; 157, Dragan Trifunovic/Shutterstock; 158, Kateryna Dyellalova/Shutterstock; 160, graham s. klotz/Shutterstock; 161, Peter Dazeley/Getty Images; 162, Hans-Joachim Roy/Shutterstock; 168, Last Resort/Getty Images; 170, aimvotalphotos/Shutterstock; 172, Mike Kemp/Getty Images; 173, Thomas Northcut/Getty Images; 175, Dino O./Shutterstock; 176, Tintan/Getty Images; 178, Pablo Eder/Shutterstock. Cover: All images provided by Shutterstock.

During the research and writing of this book, all efforts have been made to provide a wealth of choices rather than a prescribed method. As always, be sure to read the manufacturer's installation and operating instructions before you start organizing.

Index

Organize Your **Digital** Life

Aimee Baldridge

Published by the National Geographic Society
John M. Fahey, Jr., *President and Chief Executive Officer*
Gilbert M. Grosvenor, *Chairman of the Board*
Tim Kelly, *Vice President, Global Media*
John Q. Griffin, *President, Publishing*
Nina D. Hoffman, *Executive Vice President; President, Book Publishing Group*

Prepared by the Book Division
Kevin Mulroy, *Senior Vice President and Publisher*
Leah Bendavid-Val, *Director of Photography Publishing and Illustrations*
Marianne R. Koszorus, *Director of Design*
Barbara Brownell Grogan, *Executive Editor*
Elizabeth Newhouse, *Director of Travel Publishing*
Carl Mehler, *Director of Maps*

Staff for This Book
Bronwen Latimer, *Project Editor*
Melissa Farris, *Art Director*
Esther Ferington, *Text Editor*
Michele Callaghan, *Clean Reader*
Connie D. Binder, *Indexer*
Mike Horenstein, *Production Manager*
Taylor Matson, *Illustrations Specialist*
Jennifer A. Thornton, *Managing Editor*
Gary Colbert, *Production Director*
Meredith C. Wilcox, *Administrative Director, Illustrations*

Manufacturing and Quality Management
Christopher A. Liedel, *Chief Financial Officer*
Phillip L. Schlosser, *Vice President*
Chris Brown, *Technical Director*
Nicole Elliott, *Manager*
Monika Lynde, *Manager*
Rachel Faulise, *Manager*

Founded in 1888, the National Geographic Society is one of the largest nonprofit scientific and educational organizations in the world. It reaches more than 285 million people worldwide each month through its official journal, NATIONAL GEOGRAPHIC, and its four other magazines; the National Geographic Channel; television documentaries; radio programs; films; books; videos and DVDs; maps; and interactive media. National Geographic has funded more than 8,000 scientific research projects and supports an education program combating geographic illiteracy.

For more information, please call 1-800-NGS-LINE (647-5463) or write to the following address:

National Geographic Society
1145 17th Street N.W.
Washington, D.C. 20036-4688 U.S.A.

Visit us online at www.nationalgeographic.com/books

For information about special discounts for bulk purchases, please contact National Geographic Books Special Sales: ngspecsales@ngs.org

For rights or permissions inquiries, please contact National Geographic Books Subsidiary Rights: ngbookrights@ngs.org

Library of Congress Cataloging-in-Publication Data available upon request

ISBN: 978-1-4262-0334-3

Printed in United States